Automobile Steering and Suspension

John Hartley

Revised by
John Warner

W0230174

Newnes Technical Books

Published by Newnes Technical Books
an imprint of Newnes Books,
a division of The Hamlyn Publishing Group Limited,
84–88 The Centre, Feltham, Middlesex, TW13 4BH,
and distributed for them by
The Hamlyn Publishing Group Limited
Rushden, Northants, England.

First edition published by Butterworth & Co. (Publishers) Ltd., 1977
Reprinted 1979, 1981
Second edition © Newnes Books 1985

ISBN 0 408 015632

Photoset by Butterworths Litho Preparation Department
Printed in England by Mackays of Chatham Ltd

Preface

This book sweeps aside much of the mystery normally associated with suspension and steering. The basic requirements of suspension and steering are described; their main features and principles, such as roll-centre heights, spring frequencies, and roll steering are explained, together with the reasons why the various factors are important. The book should therefore appeal to enthusiasts and students, especially those studying for TEC, City and Guilds and similar courses.

Most of the popular suspension systems and geometric features are dealt with in a simple manner, and the reader can either progress through the topics or concentrate on specific areas. For the second edition, while retaining a substantial amount of John Hartley's work, I have expanded certain topics and included new material, particularly on wheels and tyres.

J.W.

Contents

1
Introduction to suspension

What is suspension?

Suspension is a system that contains some form of spring and linkage to locate and control a road wheel and its movements.

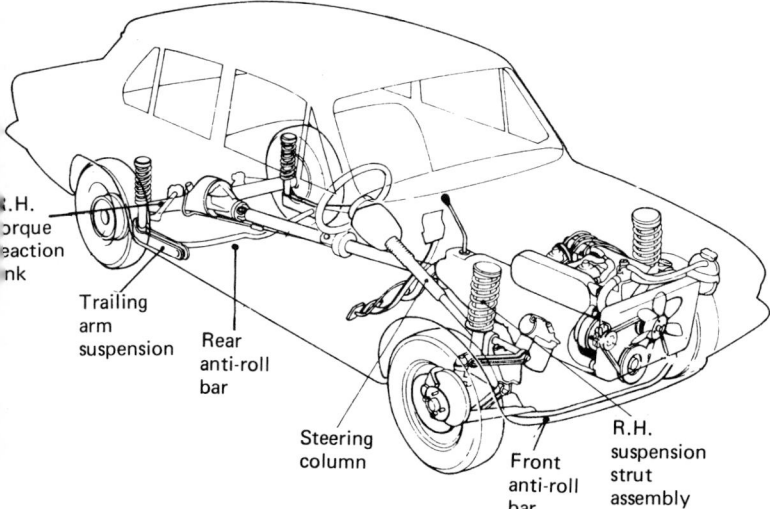

R.H.
torque
reaction
link

Trailing
arm
suspension

Rear
anti-roll
bar

Steering
column

Front
anti-roll
bar

R.H.
suspension
strut
assembly

Fig. 1. *Suspension components of a typical medium sized saloon car with rear wheel drive (RWD) employing strut type suspension at the front and trailing link with live axle at the rear*

What are the requirements of a vehicle suspension system?

It has two basic functions: (1) to isolate the vehicle, its occupants and load from road surface irregularities; (2) to maintain the road wheels (especially the driving wheels) in contact with the road surface to ensure adequate adhesion between the tyres and road for accelerating, braking and cornering.

In addition, it should limit rolling and pitching; and, for all loads, it should be self-levelling and give constant ride height and constant frequency.

Do these requirements have conflicting design features?

Yes. It is expensive and complicated to provide several of the above features, but most difficulties arise in the conflict between road holding and good ride comfort.

What are the main differences in obtaining good road holding and ride comfort?

Basically, fairly stiff (high-rate) springs are necessary to minimise body roll, adverse wheel camber changes and additional weight transfer on to outer wheels for good road holding. Soft (low-rate springs allow large wheel movements with minimum disturbance of the body to give good ride comfort.

Can a road wheel lift away from the ground?

Yes! Dampers must be fitted to restrict the effects of bounce caused by road surface humps and hollows. During hard cornering an inside wheel may lift owing to movements of the suspension linkage as the body rolls outwards.

Why does a car roll?

When cornering, centrifugal force acts on the body and attempts to push it away from the corner. This action is resisted at the road surface by the adhesion of the tyres, with the result that the body rolls about its suspension linkage.

What are the disadvantages of body roll?

Besides passenger discomfort and load shifting, there are four other disadvantages: (1) Normal suspension movements are affected as the outer springs may be fully compressed. (2) Increased weight transfer from inner to outer tyres greatly affects the attitude of the car to the cornering line. (3) Probable adverse wheel camber changes affect steering ability. (4) Body movements may change the relative positions of the roll centres and centre of gravity.

Do soft springs have any disadvantages?

Although soft springs give a good ride in most circumstances, they allow the body to roll a lot during cornering. In practice, spring rates are a compromise between the requirements of ride and handling. Excessive roll can also make the wheels adopt unfavourable angles, depending on the type of suspension fitted.

How is the softness of a spring measured?

Every spring has a *rate*, which is an indication of how much the spring is deflected by load. This is normally indicated as load per unit deflection, given in newtons per millimetre. However, when it comes to ride, the weight being carried has to be taken into account as well as the spring rate. For example, if a small mass is applied to a spring there is hardly any deflection, whereas a big mass applied to the same spring will give a big deflection. If the mass is then bounced on the spring, the small mass moves a short

3

distance very quickly, whereas the big mass moves up and down quite a long way, but slowly. The amount the mass deflects the spring is called the *static deflection*, and the rate at which the mass bounces up and down is the *natural frequency* or *periodicity* of the spring/mass system; see page 5.

Is the rate of a spring constant?

Normally yes, but it depends upon the type of spring and the requirements of the suspension system. A variable rate can be obtained, and may be necessary in order to prevent increasing load using up all the static suspension movement (Fig. 2).

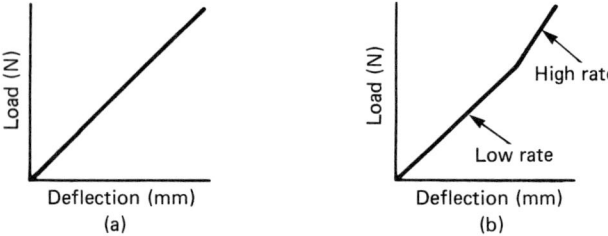

Fig. 2. *Constant rate spring (a) and variable or progressive rate spring (b)*

Which types of spring have variable rate?

Besides specially formed coil springs or the use of a compound linkage with coil springs, rubber, gas and air may be employed to provide inherent variable-rate suspension.

What other component is necessary to complete the system?

Dampers are necessary in order to limit the number of oscillations of the springs to obtain good road holding and ride. They are sometimes referred to as shock absorbers, but this is the function of a spring!

How does a damper work?

Basically a damper converts vibration energy into heat energy by restricting the flow of hydraulic fluid through valves during 'bump' and 'rebound' suspension movements. The restricting, and therefore damping effect, is usually made greater during rebound to minimise disturbance to the sprung mass (body) during bump. This is known as differential damping action.

Early dampers employed friction discs, but they were inconsistent and non-differential.

How is the static deflection found?

The static deflection, D mm, is simply the mass carried divided by the spring rate, i.e. D = mass/spring rate.

How is the natural frequency found?

This is a bit more involved, in that the natural frequency = $30/\sqrt{D}$ cycles/min, where D is the static deflection, in metres.

What are typical values?

Values vary according to the type of car. For a small saloon the static deflection is often 110–140 mm, and the equivalent natural frequency between 90 and 80 cycles/min. For a medium-size saloon the figures might be 130–180 mm, between 85 and 70 cycles/min; for a large saloon, 180–280 mm, between 70 and 55 cycles/min. With some very advanced suspensions, figures of over 280 mm or under 55 cycles/min are obtained.

Do these figures vary with load?

The actual figures for a car vary according to the load carried. For an unladen car the frequencies will normally be higher than when it contains four people and their luggage. Also, there is normally little difference in the amount of weight carried by the front springs, laden or unladen, but a big difference at the back.

Typical figures for a medium-sized saloon are:

Unladen, front	173 mm;	72 cycles/min
Laden, front	180 mm;	70 cycles/min
Unladen, rear	107 mm;	92 cycles/min
Laden, rear	140 mm;	80 cycles/min

At the front the difference is small enough to be ignored, but at the rear it makes the ride firmer unladen than when fully laden. Ideally the frequency should remain constant, irrespective of loading, and at a satisfactorily low level.

Can a constant frequency be obtained?

Only if the spring rate increases in proportion as the load it carries increases. The balance can also be improved by arranging for the increase in weight to be applied equally to the front and rear springs. In practice this is very difficult to arrange.

How is the stiffness of the spring measured in roll?

Basically, the stiffer the spring, the less the car will roll, although the spacing of the springs and the *roll-centre height* also affect the amount the car rolls. The *roll stiffness*, which is the measure of the way in which the springs resist roll, is given for independent suspensions by:

$$\text{Roll stiffness} = C_w \times t^2 \times 0.08729 \text{ Nm/deg}$$

where C_w is the spring rate *at the wheel*, known as the *wheel rate*, and t is the track of the car. For beam-axle suspensions:

$$\text{Roll stiffness} = C_s \times s^2 \times 0.08729 \text{ Nm/deg}$$

where C_s is the spring rate, and s is the *spring base*, i.e. distance between the effective positions of the left- and right-hand springs.

In simple terms this means that an increase in either the spring rate or the spring base (or track) will increase resistance to roll, and that more can be gained by increasing the spring *base* (or track) than by increasing the spring *rate*.

How can the spring base be increased?

With a beam axle the spring base is the distance between the springs (Fig. 3) and the practical width is determined by the necessity of giving clearance around the wheels and brakes. In

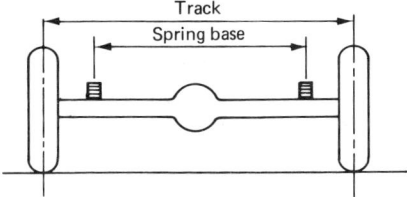

Fig. 3. *Spring base compared with track on a beam axle*

practice, this means that the spring base is about 70–80 per cent of the track, and that the wheel rate in roll is only about 50–70 per cent of the wheel rate in bump. There is little that can be done to increase the spring base with beam axles.

Is independent suspension better, then?

A better balance between ride and roll stiffness can be obtained with independent suspension, because the wheel rate in bump is the same as it is in roll. So, assuming that the linkages are equally good and that the roll centre of the independent suspension is not very low (see later), independent suspension has this one definite advantage.

7

Can roll stiffness be increased in any other way?

The addition of an *anti-roll bar* increases roll stiffness and reduces roll. An anti-roll bar is a horizontal, transverse torsion bar, normally mounted in two rubber bushes on the body. The ends are turned round to form levers, and are attached to the suspension member or axle, either directly or through drop-links. When the front wheels strike an undulation in the road so that both springs are deflected, the anti-roll bar merely rotates, so it is not deflected and exerts no force on the body. However, when the car goes round a corner, and rolls, one lever goes upwards and the other downwards, so the anti-roll bar is twisted. The bar resists, and tries to twist the car body back to the vertical position. Generally, then, the addition of an anti-roll bar reduces roll but does not affect the ride. The stiffness of an anti-roll bar is usually referred to as its *spring rate in roll*.

Does the anti-roll bar have any side effects?

Quite often only one wheel hits a bump, and in this case the anti-roll bar does alter the spring rate. The anti-roll bar twists as the wheel is raised but, since the other wheel does not move, the bar twists over its whole length (in roll, the bar is twisted from both ends, as it were, so its effective length is half the actual length). This situation, called single-wheel bump, has a higher bump rate than when both wheels move upwards together. For example, if the spring rate of the anti-roll bar is 7 N/mm and the wheel rate of the spring is 20 N/mm, then on single-wheel bump the rate is $20 + 7/2 = 23.5$ N/mm. If the proportions are of this order, the passenger will hardly notice the difference. However, if the spring rate is very low and the anti-roll bar is very stiff, a single-wheel bump will tend to rock the car, inducing what is called a roll-rock condition, which can be uncomfortable.

Can the resistance to roll be altered in any other way?

Although the roll stiffness depends purely on the spring rate in roll, the anti-roll bar and the spring base, the resistance to roll also depends on the *roll centre*. This is covered in more detail in the next chapter, but in essence the higher the roll centre the less the vehicle rolls.

Apart from the problems with roll, are there any other limitations on how soft the springing can be?

Even if roll could be kept to the minimum by the use of unusual suspension geometry and anti-roll bars, there are limitations on how soft the springing can be. The main one is the need to absorb all the energy applied to the spring by road shocks without the spring 'crashing through'. For example, let us assume that a spring has 76 mm of movement from the normal laden condition to full bump, and that the spring rate is 18 N/mm. Therefore, to compress the spring fully (or make it crash through), 1350 N must be applied by a bump in the road; the force will depend on the severity of the bump and the speed of the car, but loads of 890–1100 N can be applied quite often. Now, consider a car with a same amount of wheel travel and a very soft suspension, with spring rates of only 10 N/mm. A force of 760 N would take up all the wheel travel (10 N/mm × 76 mm = 760 N); so if a force of 890 N were applied not only would the springs be fully compressed, but a force of 130 N would be applied to the body. That would jolt the car, and also create quite a noise.

Why cannot the wheel travel be increased?

More wheel travel can be incorporated to alleviate this problem, but there are a number of difficulties. First, the wheel arches have to be made taller to allow clearance for the wheels, and this can reduce passenger or luggage space. Secondly, the springs and dampers have to be longer, and there may be little room for them (especially in low cars). Thirdly, it is difficult to arrange for the suspension to give suitable wheel control for large suspension travel.

How much wheel travel can be provided?

It is normal for there to be about 75–100 mm from normal laden to full bump; up to 150 mm is practical, especially on big cars.

How else can 'crash-through' be avoided?

The use of springs with *progressive rates* and *self-levelling systems* can allow the use of softer springs without the need for a lot of wheel travel. A spring with a progressive rate is one in which the rate increases as the spring is deflected. In the normal laden condition the rate might be 14 N/mm, after 25 mm of extra compression it might be 18 N/mm, and at full bump after 75 mm of compression it might be 25 N/mm. Thus, although the rate at the normal laden condition might be lower than in our original example, the load needed to compress the spring fully would be slightly higher.

What does a levelling system do?

It allows a car to ride at the same height irrespective of its loading. Usually it is hydraulic, and as the load increases and the rear of the car goes down, so the hydraulic system pumps the hydraulic struts up, raising the rear of the car to the original height. Equally, when someone gets out and the car rises a little on its suspension, the system removes fluid so that the car falls to its correct level.

What is the advantage of self-levelling?

The suspension has to be designed to suit the car when the driver is alone, when there are two, three or four people in it, and when there is some luggage in the boot as well. With a normal suspension, every time someone gets in the springs are compressed a little. So when the driver is alone there might be 115 mm bump travel, gradually reducing to about 75 mm when the car is laden normally, and perhaps only 50 mm when fully laden.

There must also be some *rebound travel* to allow the wheel to fall down below the normal position, for instance when the wheel goes over a pothole. Therefore, there might be 150 mm wheel travel (normally 75 mm bump and 75 mm rebound) but fully laden, when the bump travel is most needed, there would be only 50 mm travel. Now, with self-levelling there might still be 150 mm of wheel travel, but there would always be 88 mm bump travel and 62 mm rebound travel, for example. Not only does the car look better, because it always rides level, but there is effectively more suspension travel available, so softer springs can be used.

Is the amount of rebound travel important?

It is essential that the suspension has sufficient rebound travel, too little being a common fault with many earlier cars. In the first place, if the wheel is able to fall sufficiently to keep in contact with the road when it encounters a depression in the road surface, the ride is much more comfortable and there are less shock loads than if the wheel is hanging in mid-air for part of the time. But rebound travel is extremely important on cornering: when a car corners the body rolls, so as far as the suspension is concerned it is as if the inner wheel goes on rebound, and the outer wheel on bump. If there is insufficient rebound travel the inside wheel will become airborne at reasonably high corner speeds, thus reducing the cornering power. To keep some load on the inner wheel when cornering, at least 62 mm of rebound travel is needed.

2
Roll centres

Where are the roll centres of the various suspension systems, and what effect do they have on suspension geometry and vehicle handling?

The roll centres are shown in Fig. 4. With an independent suspension, the roll-centre height has a major effect on the camber change as the wheel moves up and down. Thus the parallel trailing link design has a roll centre at ground level; the wheels move vertically on bump and rebound, and so during cornering lean outwards (positive camber), parallel with the body.

If double wishbones are parallel, the roll centre is again at ground level, and the wheels move more or less vertically. If the axes of the double wishbone converge towards the centre of the car, however, the roll centre will be above the ground: the greater the convergence, the higher the roll centre. In practice with this type of suspension, cars usually have roll centres 75–150 mm above the ground, since if the roll centre is much higher the suspension behaves somewhat like a low-pivot swing axle. The higher the roll centre the more the wheel tends to lean inwards (negative camber) on bump movement, so that on roll the wheel may hardly lean outwards at all. In fact, on racing cars, the front wheels are set up with a little negative camber, and then the geometry is designed so that on roll the wheels continue to lean inwards or are upright.

With semi-trailing links the roll centre is usually a little way above ground level, and gives the wheels negative camber on bump and positive camber on rebound.

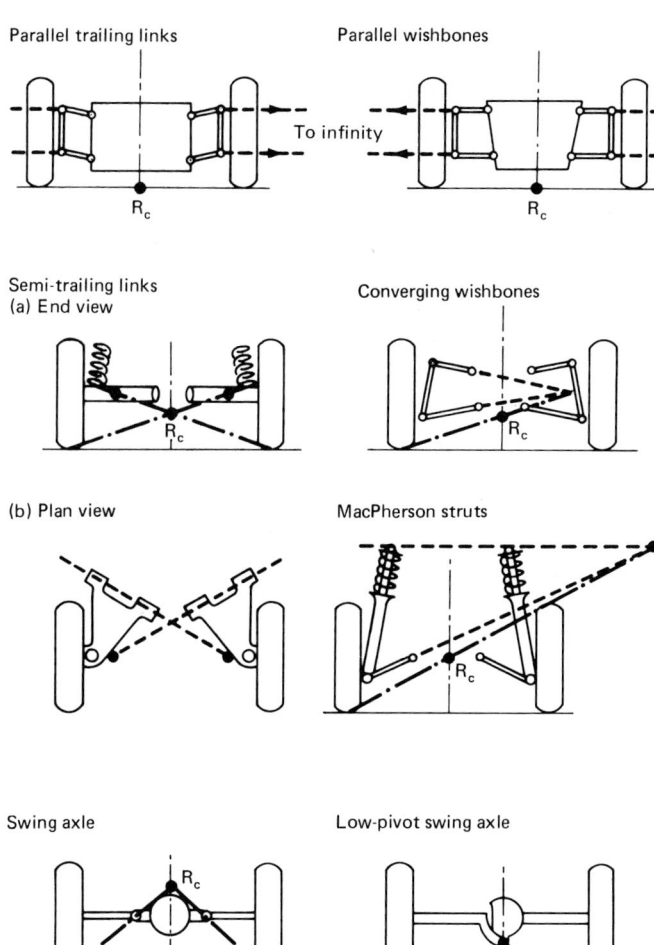

Fig. 4. *Roll-centre heights of independent suspensions*

Parallel trailing links

Parallel wishbones

To infinity

R_c

Semi-trailing links
(a) End view

R_c

Converging wishbones

R_c

(b) Plan view

MacPherson struts

R_c

Swing axle

R_c

Low-pivot swing axle

R_c

The MacPherson-strut suspension also has a roll centre a little way above ground level, and gives positive camber on rebound, but hardly any negative camber on bump.

Why does a high roll centre reduce roll?

When a vehicle corners the centrifugal force acts through the centre of gravity, and it is this that makes the vehicle roll. However, if we consider the front of the car (i.e. the front wheels, and the amount of the car they carry), then the body or portion of the body must roll about some point and that is the roll centre (Fig. 5). The force making the vehicle roll is called the *tilting moment*, which is the tilting force times its effective height. The

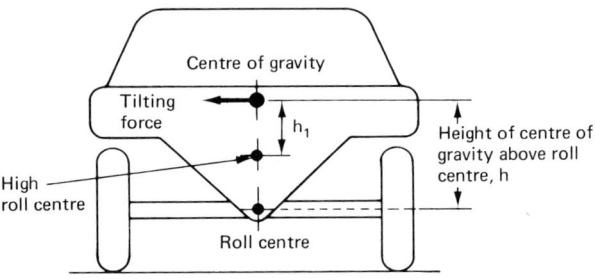

Fig. 5. *The tilting force on a body when it corners*

effective height is the distance, h, from the centre of gravity of the vehicle to the roll centre. Therefore, to reduce roll, h should be kept to the minimum. If the two were at the same height the car would not roll at all. It is for this reason that the centre of gravity of a racing car is as low as possible.

How can the various effects of roll centre height be summarised?

The most important aspect is the relationship of the front and rear roll centres to the centre of gravity. This is because the centrifugal force created during cornering acts through the centre of gravity and causes the car to roll about its roll axis – the centre line joining the front and rear roll centres.

The roll angle is proportional to the vertical height between the centre of gravity and roll axis. Basically if the centre of gravity is high and the roll axis is low the greater will be the roll angle for a given side force. Conversely, if the centre of gravity is lowered and/or the roll centre is raised, the roll angle will be reduced.

Are there any other effects associated with roll centre height?

Yes. Effects, other than body roll angle, that are greatly influenced by roll centre height are:

(1) Lateral weight transfer (during cornering).
(2) Wheel camber changes (bump rebound and roll).
(3) Gyroscopic torque.
(4) Tyre scrub.
(5) Jacking-up effect.

Besides the unpleasant effects of gyroscopic torque on steering, and tyre wear due to scrub, it is the effects of weight transfer and wheel camber change that influence the cornering characteristics of understeer, oversteer or even neutral steer.

What is the reasoning behind roll centre height and weight transfer?

Taking each of the five effects in turn, reference to Figs. 6 and 7 should illustrate clearly how weight transfer to the outer wheels can be thought of in two stages:

(1) Body roll compresses the outer springs, and therefore transfers weight directly to the outer wheels. The roll angle is determined by the height (*a*) of the centre of gravity above the roll axis.
(2) Direct weight is transferred to the outer wheels by overturning forces F_F and F_R acting through the front and

15

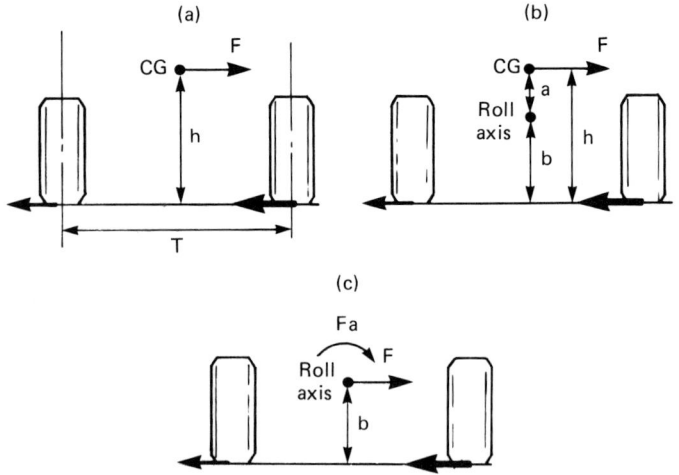

F = Centrifugal force
Fa = Rolling moment about the roll axis
CG = Centre of gravity

Front roll centre

F_F

CG

F

Fa

a

F_R

Roll axis

Rear roll centre

Fig. 6. *Relationship between centre of gravity and roll axis*

(a)

CG

F

h

T

(b)

CG

F

Roll axis

a

b

h

(c)

Fa

Roll axis

F

b

Fig. 7. *Two stages of weight transfer*

16

rear roll centres above ground level. Weight distribution and roll centre heights determine the actual amount of weight transferred.

Why think of weight transfer in two stages?

Figure 7(a) shows that total weight transfer depends upon only the height of the centre of gravity above ground level and track width, with centrifugal force F acting through the centre of gravity. Note the resisting forces at ground level through the tyres' contact patches, so that weight transfer = overturning moment (Fh).

In Fig. 7(b) the roll moment has been separated from the total moment so that the weight transfer = Fh = $F(a+b)$ = Fa + Fb.

Fa is the rolling moment about the roll axis, whilst Fb is the moment about the ground which the centrifugal force F would produce if it acted through the roll axis instead of the centre of gravity (Fig. 7(c)). If the front/rear weight distribution is known, F can be divided in the same proportions acting through the front and rear roll centres

It follows that, if one of the roll centres is raised, weight transfer will be increased at that end and reduced at the other. In other words it is a question of how the total weight transfer is shared.

How is wheel camber affected by roll centre height?

A high roll centre gives a short equivalent swing-axle length, which results in large, violent camber changes (steering kick and tyre scrub) over bumps – see Fig. 8 – and high weight transfer. On the other hand, a low roll centre with its long equivalent swing-axle length gives minimal camber change on bump but causes the wheels to lean with the body (low roll stiffness) during cornering, thus reducing their cornering power and giving low weight transfer; see Fig. 9.

O = Instantaneous centre
RC = Roll centre

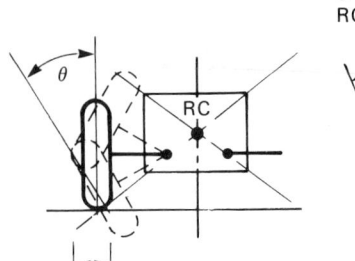

Scrub

Fig. 8. High roll centre

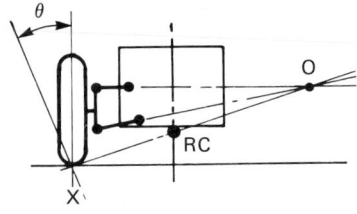

O—X = Equiv. swing-axle length

Fig. 9. Lowered roll centre

Does the tyre scrub caused by camber change have any other effects?

Scrub obviously wears the tread, and the gyroscopic reaction to the camber change is dealt with elsewhere, but a third effect is the lateral disturbance of the tyre's contact patch, which causes a side thrust and can upset the car's stability.

Which suspension layout gives the most jacking-up effect?

All independent suspensions are affected by this problem to a greater or lesser degree. The layout worst affected by jacking is the swing axle, with its high roll centre and short swing axles. It would seem therefore that a long equivalent swing-axle length with its attendant low roll centre is best, but the effects of body roll on wheel camber then become a problem. The answer is a compromise design with a roll centre height somewhere between the extremes.

18

3
Cornering and handling

What is meant by 'cornering'?

This is simply the ability of a vehicle to follow a curved path accurately whilst acted upon by lateral forces, and to retain full traction and directional control with all wheels remaining in contact with the ground.

Is 'handling' concerned with cornering?

Very much so! *Handling* is the way in which a vehicle responds to the driver's commands when making changes in direction, i.e. the ease and precision with which it is possible to change course, especially at high speed.

What are the most important factors in cornering and handling?

Some important terms are:

(1) Tyre slip angle
(2) Tyre cornering power
(3) Understeer
(4) Oversteer
(5) Neutral steer
(6) Centre of gravity
(7) Weight transfer
(8) Roll centre
(9) Roll axis
(10) Roll stiffness
(11) Swing axle effect

What is meant by 'slip angle'?

The words imply a loss of adhesion between the tyre and road surface, but what is really meant is the angle between the plane of the wheel and its actual direction of travel (Fig. 10).

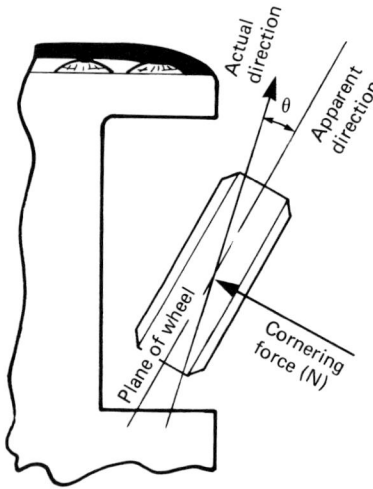

Fig. 10. Slip angle

Does the slip angle remain constant at all road speeds?

No: it increases as the speed increases. Thus although a slip angle of only, say, 5° is produced at 30 km/h, this may be increased to 10° at 80 km/h, which will alter the cornering characteristics of the car, i.e. cause the car to steer more sharply into the corner or take a wider path (depending on whether we are considering the front or rear tyres: see below).

Does the slip angle depend only on speed?

No; several factors affect the slip angle produced by a tyre, including the construction of the tyre, i.e. radial or crossply, and also its pressure, profile, wheel camber, load and speed. Note also that the slip angle generates a cornering force, at right angles to the wheel/tyre plane, which actually enables a car to turn.

What influence has the slip angle on cornering power?

If a tyre can transmit its cornering force with a small slip angle, it is said to have high cornering power.

If the slip angles are greatest at the front, the car will *understeer*, i.e. tend to run wide during cornering, but if the slip angles are greatest at the rear the car will *oversteer*, i.e. tend to turn more sharply.

How can the slip angles be made to differ between the front and rear tyres?

One of the most common methods is to employ different tyre pressures: increased pressure at the rear increases their cornering power. Other methods include increasing the vertical load of the outer wheel by the use of an anti-roll bar at whichever end it is required, and employing suspension geometry that causes the wheels to lean (called *camber*). A wheel that leans towards the centre of turn (negative camber) has a lower slip angle and hence increased cornering power.

Can the driver's actions affect a tyre's cornering power?

By use of the accelerator and brakes a driver can increase or decrease the cornering power. The cornering power of rear driving wheels decreases when accelerating and increases slightly when braking, owing to the weight transfer effect.

How does the driver know whether the car is understeering or oversteering?

When travelling along a straight road an understeering car will exhibit an automatic self-cancelling effect to any side thrusts, e.g. road camber and wind gusts. This is due to the centrifugal force generated as it tends to steer slightly away from the side thrust when the front slip angles are greater than those at the rear. During cornering an increased amount of steering lock has to be applied, owing to the tendency to run wide when understeering.

The opposite effects are produced when oversteering, i.e. the car will tend to steer towards the side thrust so that the centrifugal force generated will now add to the effect and cause even greater disturbance. During cornering the steering lock must quickly be reduced to prevent the car mounting the kerb, and perhaps also to prevent a spin. Understeer inparts stability and so is safer for normal motoring.

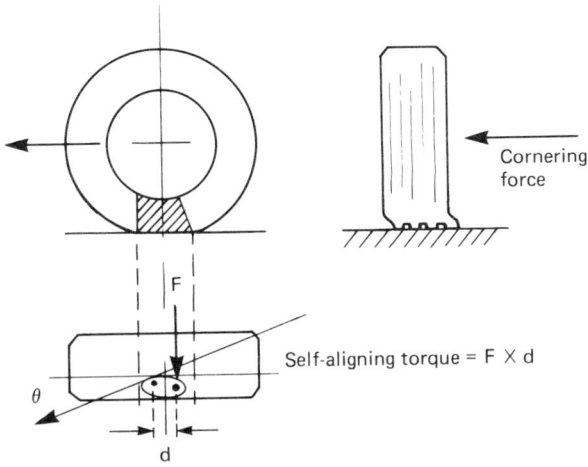

Fig. 11. *Self-aligning torque*

Does the cornering force affect the shape of a tyre at the road surface?

A pneumatic tyre deflects and becomes distorted during cornering. This sets up the cornering force, which acts slightly to the rear of the tyre's contact patch or footprint and results in a turning moment generally known as self-aligning torque (Fig. 11). It is this self-aligning torque (plus the castor angle: see page 79) that gives the driver 'feel' and is therefore a measure of the force required to steer the car.

How much self-aligning torque is provided?

The actual degree or value of self-aligning torque depends upon several factors, such as tyre design and pressure; it gradually increases as the slip angle increases, until it reaches a maximum and then falls off. A skid is then imminent and control is lost. Too much self-aligning torque produces heaviness, which is tiring to the driver.

What factors influence the cornering power of a tyre?

(1) *Inflation pressure.* Cornering power increases with inflation pressure to the limits of comfort and wheel hop.
(2) *Vertical load.* Increased vertical load decreases cornering power, especially at the larger slip angles.
(3) *Wheel camber.* A tyre that leans towards the centre of turn has a higher cornering power (negative camber) than one that leans away (positive camber).
(4) *Tyre construction.* A radial-ply tyre has a higher cornering power than a crossply. Steel-braced radials are higher than textile-braced.
(5) *Tyre profile.* Owing to the wider tyre section, lower-profile tyres have a higher cornering power.

(6) *Wheel rim width.* Wider rims increase cornering power but introduce increased side-wall stresses.

(7) *Traction and braking.* Cornering power decreases under traction and braking.

Cornering power is largely a matter of tread distortion (slip angle) for a given load. Less distortion gives increased cornering power.

4
Anti-roll bars

What is an anti-roll bar?

It is a transversely mounted torsion bar in two rubber-bushed bearings and connected by drop links to the front suspension lower arm. Additional anti-roll bars may be fitted to the rear suspension. See Fig. 12.

What does an anti-roll bar do?

It improves road holding, not only by resisting body roll but also by increasing roll stiffness, which tends to maintain the more heavily loaded outer wheels more upright, thus increasing the cornering power of the wheels at whichever end of the car the anti-roll bar is fitted. If a wheel leans outwards it develops less cornering power, and vice versa.

Is weight transfer affected by the use of anti-roll bars?

The *total* weight transfer for the whole car depends upon the height of the centre of gravity above ground level and upon the width of track (Fig. 13), so anything that *increases* the weight transfer at one end *decreases* it at the other and vice versa.

The use of an anti-roll bar directly increases weight transfer at the end to which it is fitted by increasing roll stiffness; thus the *total* weight transfer is shared according to the roll stiffness of the front and rear suspension.

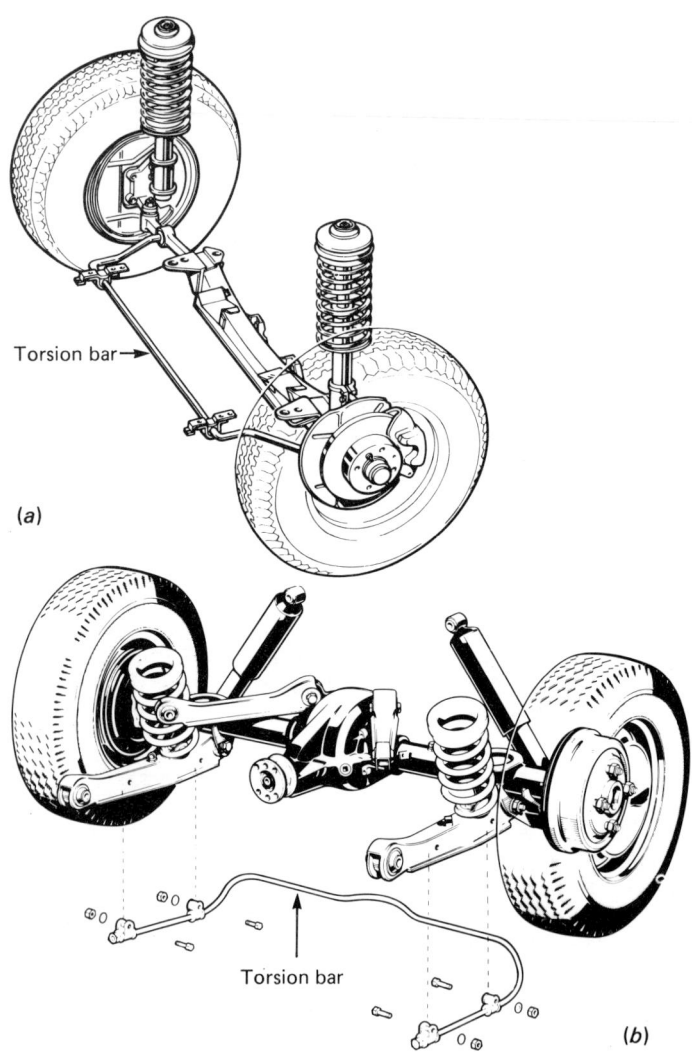

Torsion bar→

(a)

Torsion bar

(b)

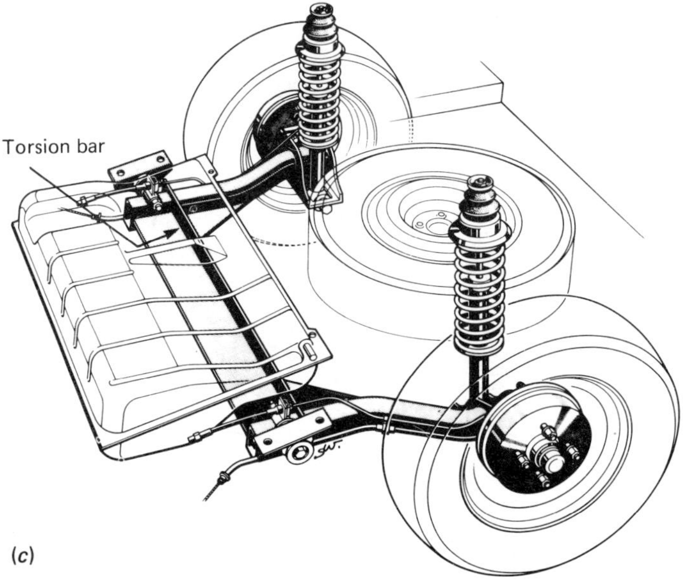

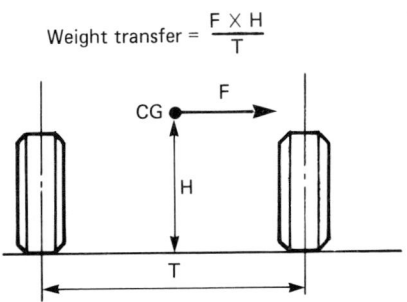

Torsion bar

(c)

Fig. 12. Three typical anti-roll bar installations on (a) Fiat 131 front, (b) Ford Cortina rear, (c) Volkswagen Golf rear

Weight transfer = $\dfrac{F \times H}{T}$

CG● ——→ F

H

T

Fig. 13. Total weight transfer

27

What is so important about the way weight transfer is shared?

The cornering force of a tyre depends upon several factors, including the vertical load it carries. It can be seen from Fig. 14 that the slip angle increases as the tyre load increases, for a given

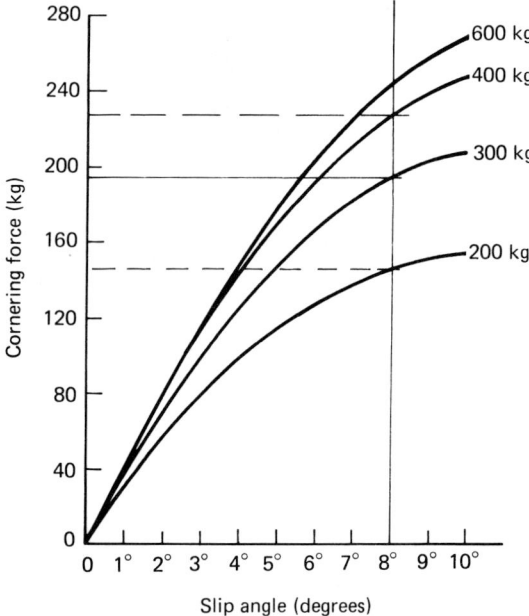

Fig. 14. *Consider a pair of wheels supporting a weight of 600 kg. With the weight divided equally between the wheels, at a slip angle of 8° a cornering force of 392 kg (2 × 196 kg) would be produced. But with a weight transfer of 100 kg the relevant cornering forces of the two tyres (at an 8° slip angle) would be 228 and 145 kg, a total of 373 kg. To maintain cornering power the wheels would have to be turned to a 9° slip angle (242 + 150 = 392 kg), and 1° more at the wheels means about 15° more at the steering wheel. So increased weight transfer on a pair of wheels reduces their cornering power*

cornering force. Thus increasing the vertical load acting down on a tyre increases its slip angle, which reduces its cornering power. This effect produces either understeer or oversteer behaviour according to which wheels are affected, front or rear.

It is generally found that if an anti-roll bar is fitted to the front suspension the car will understeer because of increased mean slip angles at the front, and will oversteer if one is added only to the back.

If anti-roll bars are fitted to the front and rear suspensions, how can changes be made to the car's handling?

If the ratio between the front and rear anti-roll bar effects is to be maintained, several other factors can be varied that will, for instance, reduce an original understeer – for example, lowering the front suspension roll centre, increasing the front-wheel negative camber, reducing the rear-wheel negative camber, or increasing the front-wheel toe-in.

If it is thought desirable, a thinner front anti-roll bar can be fitted to reduce understeer and vice versa, but other undesirable factors can then be a problem.

Does an anti-roll bar affect pitching?

No, because the bar needs to be twisted, as a torsion bar, for it to be effective.

5
Suspension systems

Which form of suspension was most popular before independent suspension was introduced?

Leaf springs of some form or other were used on nearly all the early cars, with semi-elliptic springs (Fig. 15) the most popular. Some rear axles were carried on quarter-elliptic springs (Fig. 16), but these were inevitably very stiff. To overcome this, semi-elliptic springs were sometimes used in cantilever form, the axle being mounted on the ends of the springs, which were achored on the chassis; midway along the spring was a bearing pad on the chassis. A single semi-elliptic spring was mounted *transversely* (Fig. 17) on the early English-built Fords and on the Austin 7s. In this case there was a shackle at each end of the spring, and so the lateral location of the axle depends on the flexibility of the shackles! Longitudinal location was afforded by a pair of radius rods. Although these alternative schemes tended to be cheap, they all had drawbacks compared with the semi-elliptic suspension in which a pair of springs were installed *longitudinally*.

What is wrong with semi-elliptic suspension?

The conventional beam axle with semi-elliptic springs is adequate so long as the passengers are prepared to put up with a hard ride, and so long as the car does not have too much power. To make the ride softer, springs with a lower spring rate must be used;

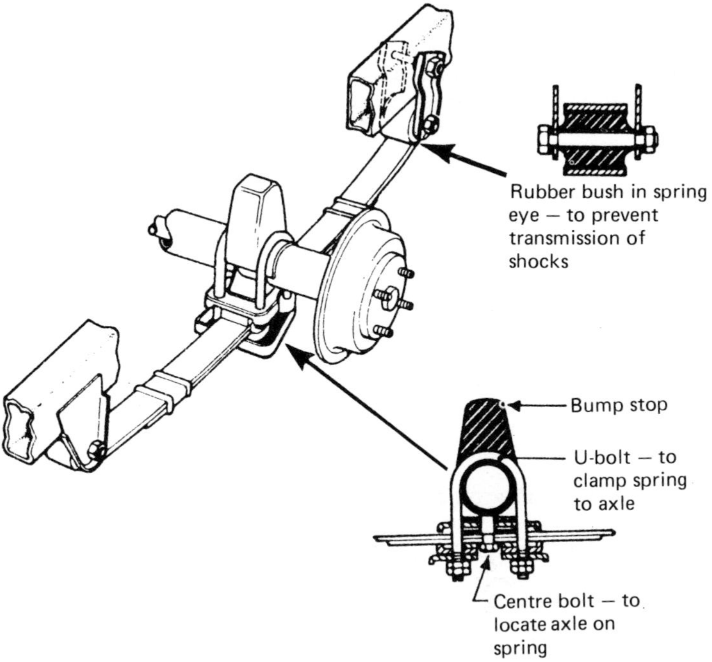

Rubber bush in spring
eye — to prevent
transmission of
shocks

Bump stop

U-bolt — to
clamp spring
to axle

Centre bolt — to
locate axle on
spring

Fig. 15. *A laminated or leaf spring is a cheap system which also locates the axle*

Fig. 16. *Quarter-elliptic leaf spring*

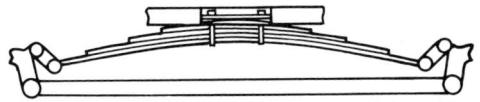

Fig. 17. *Transverse semi-elliptic leaf spring*

31

these will be longer than stiff springs, and inevitably more flexible. As a result, when the power is applied rapidly or the brakes are applied firmly, the springs 'wind up' as shown in Fig. 18, and the wheel may hop or patter and become out of control.

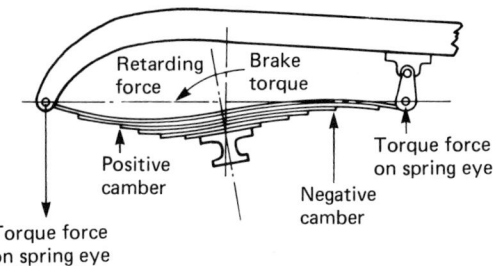

Fig. 18. *Semi-elliptic suspension may 'wind up'*

Are there any other defects with leaf springs?

Another problem is inter-leaf friction. When the spring is deflected, each blade moves slightly relative to the one next to it; because they are held against one another the friction resists suspension movement, and exerts a form of damping. Unfortunately, the amount of damping depends on whether the blades have rusted or not, whether they are greasy or the gaps filled with dust and dirt, or even whether it is wet or dry.

To some extent these problems can be alleviated, and are where semi-elliptic springs are used on modern cars. Plastics interlayers can be inserted between the blades to reduce friction. The spring can even be formed as a single blade, thicker at the middle than at the ends, in order to eliminate the inter-leaf friction altogether.

Can axle location be improved?

Although these changes reduce friction, and therefore improve the ride, the use of soft springs means that the axle is not located

32

very precisely. The solution here is to add some radius rods, which pivot on brackets above the axle, and run forwards to pivots on the body. In an ingenious solution on some Ford cars, and copied by some others, the lever arms of an anti-roll bar act as radius rods, so the anti-roll bar does two jobs (Fig. 19).

Fig. 19. *The anti-roll bar can be mounted so that its levers act as radius rods and help locate the axle (Ford Escort)*

Can anything be done without adding radius rods?

Some improvement to the longitudinal location of the axle can be effected by mounting the springs asymmetrically. That is, instead of the axle being midway along the spring, it is nearer the front anchorage, as in Fig. 19. For example, in a spring 1250 mm long the distance from the front anchorage to the centre-line of the axle would be about 500 mm. Effectively, therefore, the front portion of the spring is stiffer and can withstand the braking and acceleration forces better, but the spring rate is still fairly low. All other improvements involve the use of extra links.

Why are laminated spring suspension systems rarely used for passenger cars now?

This type of spring has several disadvantages, including poor energy storage for its weight, spring-blade inter-leaf friction, and limited range of suspension movement.

What other types of spring can be used?

There are several different forms of springing, including helical coil, torsion bar, rubber, fluid–gas, and pneumatic.

What are the special features of a helical coil?

This is very popular for front and rear suspension. It has a high energy storage factor and no inter-leaf friction, but additional linkage for axle location and reaction to acceleration and braking torque. It does, however, require increased vertical space (see Fig. 20).

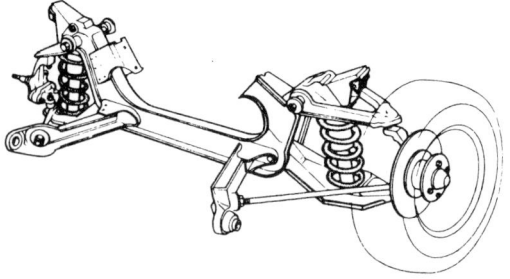

Fig. 20. *Double transverse link front suspension with helical springs*

What are the advantages of a torsion bar?

It is not as popular as the helical coil spring but possesses even greater energy storage. It is usually mounted longitudinally under the floor, and again has minimal internal friction but requires additional linkage; see Fig. 21. An interesting feature is the ability to adjust the 'ride height'.

Torsion bars can be mounted transversely when used with a trailing-link system as in the Volkswagen Beetle.

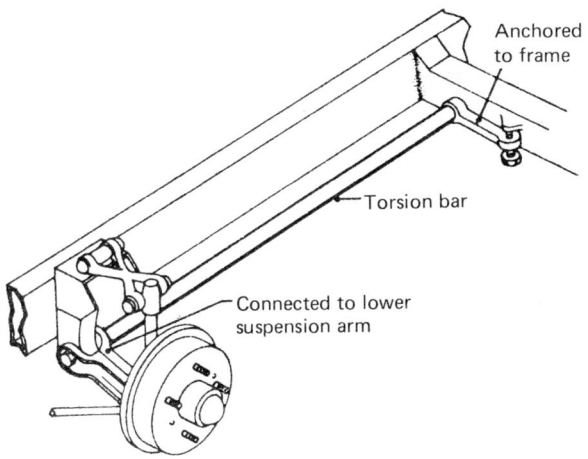

Anchored
to frame

Torsion bar

Connected to lower
suspension arm

Fig. 21. *Torsion bar*

Are rubber springs effective?

A rubber spring has the highest energy storage capacity of all springing materials and also reduces noise transmission. It is also very light and has inherent damping qualities, owing to its internal friction. See Fig. 22.

By designing their fixing brackets for loading them in shear and then changing to compression, rubber springs can easily be given a *variable rate* or *rising rate*, i.e. a soft reaction to small wheel movements but becoming harder as the suspension deflects; see later.

Rubber springs are popular for heavy commercial vehicle and trailer bogies. Two problems are their tendency to 'creep' or 'settle', which lowers the ride height early in service, and their tendency to stiffen up in cold weather.

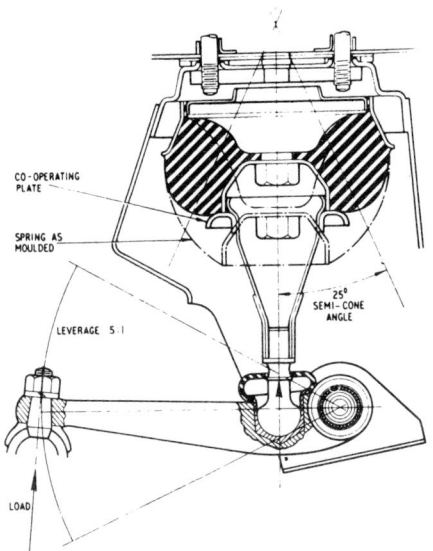

Fig. 22. *Moulton rubber cone spring installation (BL Mini)*

What fluid–gas installations are there?

Fluid–gas springing was pioneered by the Citroën company in 1953 on the rear suspensions of six-cylinder cars. They used high-pressure (16.8 MN/m^2) mineral-oil hydraulic fluid and nitrogen gas. A self-levelling arrangement was also used.

In 1973 British Leyland introduced the Hydragas system (Fig. 23), which interconnects the front and rear suspension units. The

36

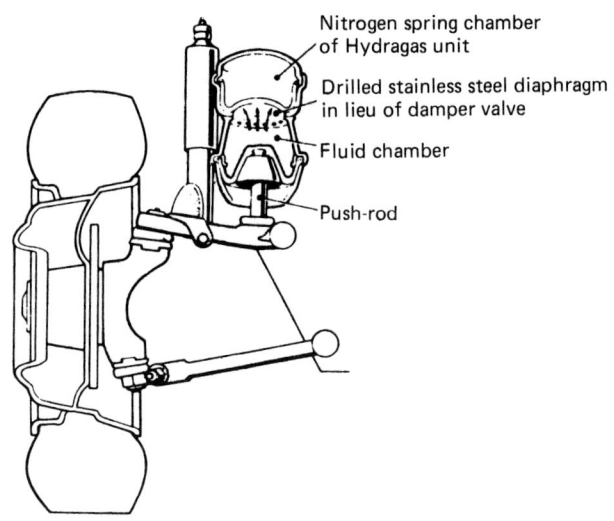

Nitrogen spring chamber
of Hydragas unit

Drilled stainless steel diaphragm
in lieu of damper valve

Fluid chamber

Push-rod

Fig. 23. *Hydragas installation in Austin Allegro*

fluid used is a mixture of water, alcohol and corrosion inhibitor. Later versions (such as on the Metro) do not have the front and rear springs interconnected.

How is a gas suspension contained?

If a special gas is used as the springing medium, as on the Citroën and British Leyland Hydragas suspensions, it is enclosed in a completely sealed chamber. A diaphragm separates the gas from a fluid, which transfers the movement to the wheel. In the Citroën design, self-levelling valves admit and exhaust fluid from the struts as needed. Should any leakage of gas take place, the levelling valve pumps the system up to the normal height.

How is this done on the Hydragas suspension?

On the Hydragas suspension there is no self-levelling system, but the front and rear suspension are interconnected. As a result the car rides relatively flat. If any gas should leak, the pressure of the fluid can be increased to retain the normal ride height.

How near the ideal suspension are these gas systems?

The Citroën system has very low-frequency suspension, so the ride is very good, while large anti-roll bars are used to limit roll (which is still substantial). The Hydragas suspension, on the other hand, has been designed to give little roll without the need for anti-roll bars, so the suspension frequencies are a good deal higher. There is no doubt, though, that hydropneumatic suspensions, as these are, are the nearest approach to the ideal suspension that is in use.

Are pneumatic (air) springs widely used?

Air springs are not commonly used for passenger cars because of the high cost, complex air system and bulk. But they are widely used on commercial vehicle tractor units, semi-trailers and large drawbar trailers, all of which vary their ride heights when operating between laden and unladen conditions.

The very high quality of ride obtainable with air suspension is especially attractive for use on coaches and is used extensively on the Continent and in the USA.

What components are normally required for an air suspension?

The air spring itself takes the form of a diaphragm, usually fixed to the chassis at the top, with its lower face resting on a conical piston. An air compressor is needed to supply air to the springs, usually at about $420 \, kN/m^2$. There must also be some valves in the system to allow air to be fed to the springs, or exhausted from them when needed. These are normally the self-levelling valves, and they are actuated by links connected to the suspension. If independent suspension is used, the link connects to the centre of an anti-roll bar, so that it is not affected by roll, but when the vehicle goes down on the springs the valve is opened to allow air to be delivered to the springs, and vice versa. To prevent this happening every time the vehicle goes over a bump, a device (a time switch, or a damper) is usually built in to allow the valve to open only when the suspension has settled down.

Are hydro-pneumatic systems more complicated?

There are more systems and more components in the self-levelling hydro-pneumatic suspension. There are the spring/strut units, the levelling valves, the pump, and a number of one-way valves. There is also a hydraulic accumulator, which stores a reserve of energy. Extreme care is needed when dealing with these systems, since they operate on high pressures. The circuits are so complex that complete data are needed on the system before any repair work is undertaken.

Variable-rate springs

Which types of spring have variable rates?

If a variable rate is desired, air, gas, rubber, or specially formed coil springs are used. Alternatively, a compound linkage can be used to give a constant-rate spring variable rates.

How is the linkage arranged to give variable rates?

To obtain variable rates by means of the linkage it is necessary for the movement of the spring to vary relative to that of the wheel. For example, if during the first inch of travel from rebound of the wheel the spring is deflected by 13 mm, then in the last inch of wheel travel the spring deflection would need to be more than this

– say 20 mm. In this case the rate would increase by just over 100 per cent from rebound to full bump: from, say, 18 N/mm to 40 N/mm. The relationship between wheel movement and spring movement depends on the linkage used and on the effective angles of those links. If the suspension arm is horizontal, and the spring inclined at about 30° to the vertical at full rebound, then as the spring is deflected so its angle will increase. As a result the movement of the spring relative to that of the wheel will increase with deflection, effectively increasing the spring rate. If this progressive geometry is applied to another linkage with similar characteristics, a substantial change in rate can be obtained. This system is used on some racing cars.

How can a coil spring be designed to give a variable rate?

In a constant-rate coil spring the pitch of each coil remains constant, i.e. there is an equal gap between each coil. A variable rate can be obtained if the coils are wound with a variable pitch: close at one end, and gradually opening out along the spring. In the first part of compression the coils at one end close up solid, so that the spring effectively has fewer coils, and thus the rate increases. A similar effect can be gained by grinding the bar so that it is tapered before winding the coil, or by a combination of both methods.

How much can the rate be made to vary?

If the coil spring has variable pitch and is taper-ground, the rate can increase 2–3 times. Between rebound and normal laden height the rate can increase by 1.5–1.8 times, and by full bump by up to 3 times.

Can other springs be given such big variations in rate?

Normally rubber springs give relatively small increases in rate as they are deflected. Although quite large variations are possible, these are usually ruled out by high stresses leading to a shorter life. Air or gas springs can be designed to have highly progressive rates, and this usually results in a suspension with constant frequency on cars.

How is the rate of the air/gas spring increased?

In a gas spring, the spring rate $C = kPA^2/V$, where k is a constant, P is pressure, A is the effective area, and V is the volume of the gas. As the spring is compressed the pressure increases and the volume reduces, so that the rate increases. By the use of a flexible diaphragm that rolls on to the piston as the gas is compressed, the change in rate can be tailored to suit any requirement.

Constant-rate springs

Which types of spring have constant rates?

Leaf springs, torsion bars and coil springs all have constant spring rates. Some of these *can* be formed or installed to give progressive rates (see above), but generally they are installed with constant rates – so that, for example, the spring rate is 18 N/mm throughout the suspension travel.

What happens when springs fail?

Occasionally springs are overstressed so that they break. In the case of a leaf spring one blade breaks, and a small change in ride height may be the only clue; with a torsion bar the suspension will go right down on to its bump stop, so the fault will be obvious. With a coil spring the chances are that the lower half of the spring will support the upper half, so the ride height will not be reduced by very much. However, the spring will rattle and the ride will be worse, with the car tending to crash through to the bump stop more often than usual.

6
Location of live axles

What are the advantages of a live axle?

Its relatively low cost and robust design suits a competitively priced saloon car with front engine and rear wheel drive, especially if helical-coil suspension is used. (See *Questions & Answers on Automobile Transmission Systems* for a description of live axles.)

What are its disadvantages?

(1) It has a high unsprung weight as a result of its malleable iron casing and alloy steel final drive, differential and drive shafts, etc. (2) The rigid connection of the road wheels means that when one wheel rises or falls over a bump the other wheel is also affected. This causes gyroscopic reactions and camber changes. (3) The torque reaction arising from the propeller shaft presses the left-hand wheel down whilst attempting to lift the right-hand wheel during acceleration, thus leading to wheel spin and 'axle tramp'. (4) Sufficient body clearance must be given to allow for the vertical movements of the axle.

How can a live axle be located when coil spring suspension is used?

There are several systems now in use, and they can be classified by the number of locating links used (Fig. 24):

(1) Four-link system with Panhard rod.
(2) Four-link system with Watt's linkage.
(3) Four-link system with semi-trailing upper links.
(4) Two-link system with A-bracket.
(5) Torque tube with Panhard rod or Watt's linkage.

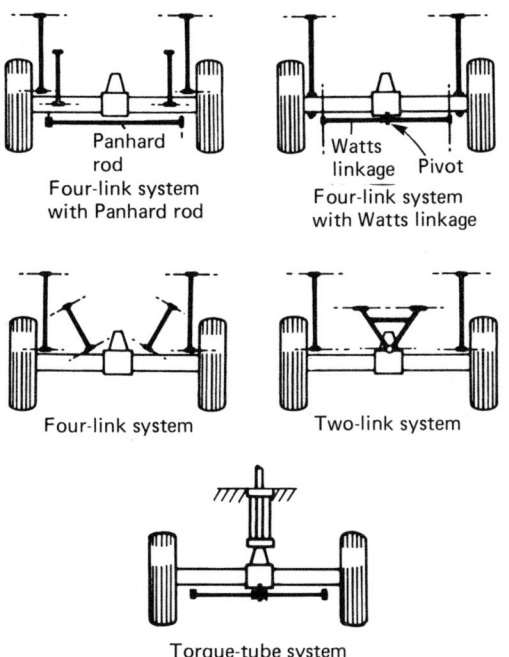

Fig. 24. *Identification of live-axle location systems*

What are the features of the five axle-locating systems?

(1) The *four-link system with Panhard rod* gives good isolation from road noise, owing to the large number of rubber bushes used

43

and is an effective means of suspension control. Its wide usage includes the Fiat 131, Mazda 626, Vauxhall Carlton and Volvo 244/264.

(2) The four-link system with Watt's linkage is similar except that a Watt's linkage replaces the Panhard rod. The Watt's linkage gives almost pure vertical axle movement but takes up more space and of course requires an extra chassis pick-up point. It is more complicated and is therefore move expensive. It is fitted on the Mazda RX-7 and Reliant Scimitar.

(3) The four-link system is the most simply located live-axle. It employs semi-trailing upper links to provide lateral location and to act as torque arms. Although relatively inexpensive it requires careful tuning of the compliance of its locating bushes to allow freedom of axle movement without road-noise transmission. It can be found on the Colt Sigma, Datsun Sunny, Ford Cortina, Fiat 132, Talbot Avenger and others.

(4) The two-link system with A-bracket is found on the Renault. It is really a variation on the previous theme, the upper two links being replaced by an A-bracket. Accommodation and high bearing loads can be a problem.

(5) With the torque tube system, reaction is resisted by a rigid tube extending forwards from the differential nose-piece. The front end of the tube pivots in a bracket secured to the body.

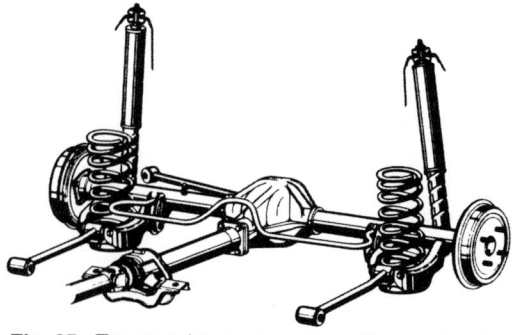

Fig. 25. *Torque-tube suspension with a pair of radius rods and a Panhard rod*

Trailing links give fore-and-aft location, while lateral location is by either a Panhard rod or Watt's linkage. This system gives better axle location, reduced drive-line noise and a lower floor line. It is found with a Panhard rod on the Opel Manta, Peugeot 504, Vauxhall Chevette and Cavalier. With a Watt's linkage it is on the Rover 2300, 2600 and 3500.

How does a Watt's linkage work?

A Watt's linkage can be employed for both lateral and fore-and-aft location of axles, ensuring they rise and fall in a straight line. See Fig. 26. The bars of the Watt's linkage, by acting as struts or tie rods with forces equal to those transmitted by the wheels to the axle, prevent any lateral movement of the body. However, it is an expensive arrangement and difficult to accommodate.

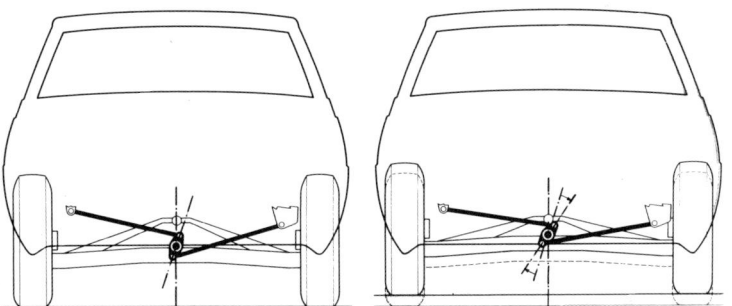

Fig. 26. Action of a Watt's linkage as the suspension deflects

Is there any other kind of rear axle?

The De Dion axle, which is known as a beam-type dead axle, is usually located by two trailing arms and either a Panhard rod or a Watt's linkage. It is used (with a Watt's linkage) by Alfa Romeo and Rover. See Fig. 27.

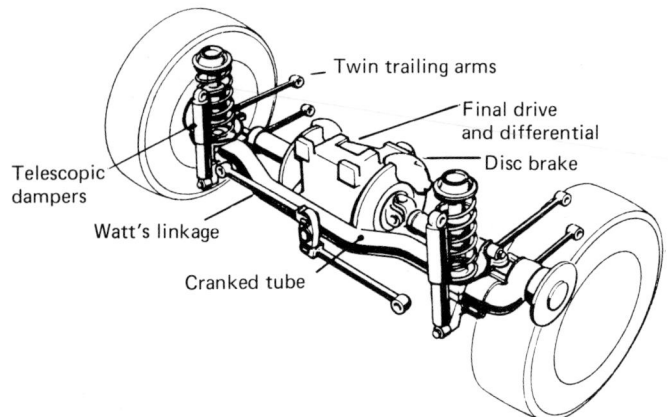

Fig. 27. *The De Dion axle is non-independent and a dead axle; note the use of inboard disc brakes*

What type of cars have used the De Dion axle?

Mainly racing and high-quality sports cars, where the need to maintain the outside rear wheel upright when cornering was paramount; this was difficult to obtain with early designs of independent suspension. The problem of one wheel on bump disturbing the opposite wheel is a disadvantage.

What are the advantages of a De Dion axle?

The main advantage is the great reduction of unsprung weight, owing to the location of the final drive and differential unit on the body. It also permits the use of inboard brakes, with further decreases in unsprung weight, which of course improves road-holding and ride comfort.

Additional advantages are the facts that propeller shaft torque is reacted by the body, and the wheels remain parallel to each other at all times.

7
Independent suspension

What are the advantages of independent front suspension (IFS)?

There are four basic advantages:

(1) *Improved steering precision* because the wheel movements are not linked from side to side, there is reduced gyroscopic reaction, and the wheel travel path is more accurately controlled.
(2) *Better road holding* through the use of softer springs and wider spring-base, aided by reduced unsprung weight.
(3) *Increased ride comfort* because of the softer springs and reduced unsprung weight. Coil springs and torsion bars are used for suspension purposes only, while radius arms and torque bars locate the axle units.
(4) *Increased passenger accommodation* arising from the ability to position the engine further forward and lower, owing to the removal of the solid axle beam which required clearance for its vertical movements. A lower bonnet line is also obtainable.

Can the same be said for independent rear suspension (IRS)?

Yes, the majority of IFS advantages can be applied to IRS, but the reduction in unsprung weight of the rear axle by about 50 per cent is especially important to the improvements in ride comfort, road holding and traction.

Passenger space is further increased by the absence of propeller shaft and final drive assembly requiring body clearance for their vertical movements with the suspension action.

What are the most common IFS layouts?

The two most popular systems are: (1) unequal double transverse links, and (2) single transverse link and strut (MacPherson).

What are features of the double-link system?

The hub and wheel assembly is carried between two triangulated links or wishbones, so designed to cope with braking torque reaction. The links may be parallel to each other or inclined to raise the front roll centre.

The links are of different length (Fig. 28). When the lower link is made longer than the top link, the effect of body roll on wheel camber is reduced whilst at the same time maintaining the track constant (Fig. 29).

Equal-length links are no longer used because they cause two further problems: wheel camber is directly related to body roll, which gives the more heavily loaded outer wheel positive camber and therefore reduces its cornering power (Fig. 30); also increased tyre wear results from the constantly changing track.

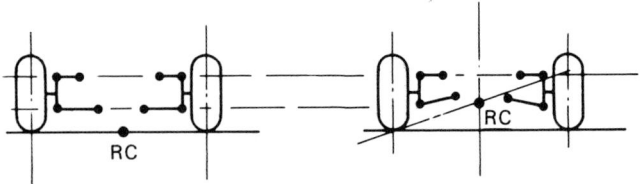

RC = Roll centre

Fig. 28. Double-link IFS: (a) parallel links give a roll centre at ground level; (b) inclined links give a raised roll centre

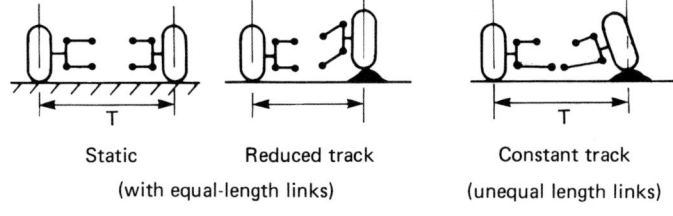

Static Reduced track

(with equal-length links)

Constant track

(unequal length links)

Fig. 29. *Effect of link length on track*

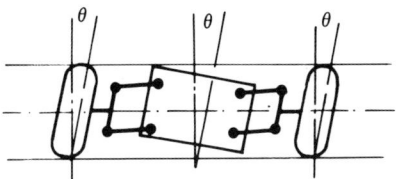

Fig. 30. *Effect of body roll on camber*

What are the features of the MacPherson strut, single transverse link IFS system?

Designed by E. S. MacPherson for Ford around 1945, this strut-type suspension consists of telescopic damper and helical coil spring connected between the body structure and single lower link, usually referred to as a track control arm. Some means of transmitting the fore-and-aft forces to the body structure is necessary, and a tie rod at 45° in plan view is used for this purpose. (It may well be the splayed ends of the anti-roll bar as in Fig. 31.)

Does the MacPherson strut suspension have any special advantages?

This simple form of suspension layout is ideally suited to front-wheel drive cars, owing to the absence of any upper links.

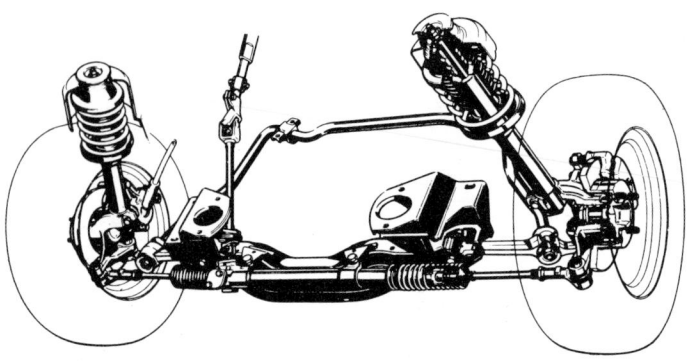

Fig. 31. *MacPherson strut suspension layout on the Ford Sierra*

With its wide mounting points it reduces the stress concentrations on the body structure. It is quite easily accommodated within the integral body construction of the modern motor car, and allows more space for maintenance and repairs around the engine assembly.

What disadvantages are found with this form of suspension layout?

There are frictional forces within the telescopic guide rod, owing to cornering and braking loads, and a sensitivity to out-of-balance forces on the roadwheels.

What other suspensions may be used at the front?

There are several alterntive layouts, such as the trailing link, swing axle and sliding pillar. The trailing link (used in the VW Beetle and Aston Martin DB2) usually has two links with torsion bar suspension, but coil springs can be used. Early Vauxhalls used a single arm in conjunction with torsion bars, torsion-bar springs and hydraulic damping.

The swing axle (as in the Hillman Imp) has two swinging half-axles pivoted at or near the centre of the car when viewed from the front. Triangulated arms, or some other form of fore-and-aft linkage, are used.

The sliding pillar (Morgan, Lancia) has roadwheels mounted on vertical sliding members acting on coil springs and controlled by hydraulic dampers.

What are the merits of the various systems?

The swing axle is the simplest and therefore the cheapest suspension, but it cannot be designed easily to prevent the transmission of noise to the car; nor is its geometry very good. The twin trailing link design has little to commend it; it is quite complicated, and although the wheels rise and fall vertically over bumps, they lean outwards on corners, which is undesirable. Trailing links are quite expensive and heavy because they have to be designed to withstand high lateral forces, which tend to bend the links. However, they do give a reasonable installation for transverse torsion bars.

The beauty of the MacPherson strut suspension is that it is very simple, with only one link and with the spring and damper incorporated in the strut, and the pivots can be very widely spaced so that they carry small loads. The geometry is not ideal, but is quite good in practice. The double link system can be designed with quite widely spaced pivots, and it can easily be mounted on a sub-frame, which can both facilitate manufacture and also help isolate the body from road noise – as it does very well indeed on the Jaguar cars. The links can be arranged to give wide variations in geometry. The cost of the suspension also varies widely according to the design.

What types of independent suspension are used at the rear?

Again, there are many types of suspension employed, showing how difficult it is to design a system that not only improves handling and ride but does so at a reasonable cost. The choice of suspension layout is also influenced by whether the drive is by way of the rear wheels (RWD) or front wheels (FWD).

The principal types are: swing axles, trailing arm, semi-trailing arm, unqual-length transverse links, and transverse link and strut.

Why has it taken so long for IRS to become accepted?

Basically the use of a live beam axle at the rear is not such a problem as when used at the front. It is a simple structure, but it does carry a high unsprung weight.

With front-wheel drive becoming the accepted method of obtaining tractive effort, the use of IRS is not so urgent, especially as the dead beam axle is relatively light and requires only simple locating links.

How does swing axle suspension work?

The layout is very simple and economical to manufacture, mainly due to use of one universal joint and no sliding coupling on each half-shaft. Note the use of a transverse semi-elliptic leaf spring to locate the wheel uprights (Fig. 32).

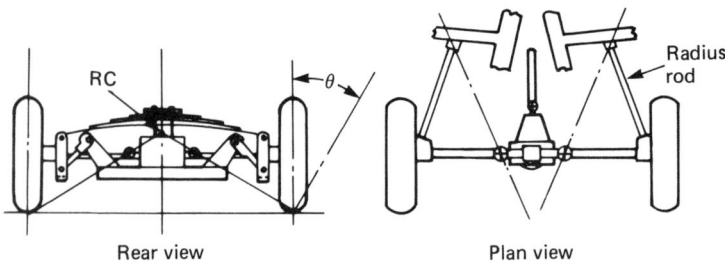

Rear view   Plan view

Fig. 32. Swing axle suspension, rear (left) and plan (right) views

A disadvantage with swing axles is the jacking-up effect when cornering. This causes the more heavily loaded outer wheel to lean outwards with increasing positive camber, thus reducing the tyre's cornering power and increasing rear slip angles (oversteer effect). Eventually the jacking effect may cause the outer wheel to tuck under, with resultant loss of adhesion.

The short swing axles give a high roll centre, which increases the lateral weight transfer and wheel camber changes coupled with a large tyre scrub-angle (θ). Mercedes adopted a low-pivot arrangement to overcome these problems.

How does trailing arm suspension compare?

This is quite a popular suspension layout, used by several manufacturers: British Leyland, Colt, Citroën, Nissan, Peugeot, Renault, Talbot. The arrangement is simple, compact and light but suffers the problem of the wheels leaning over at the same angle as the body when cornering, with loss of cornering power. See Fig. 33.

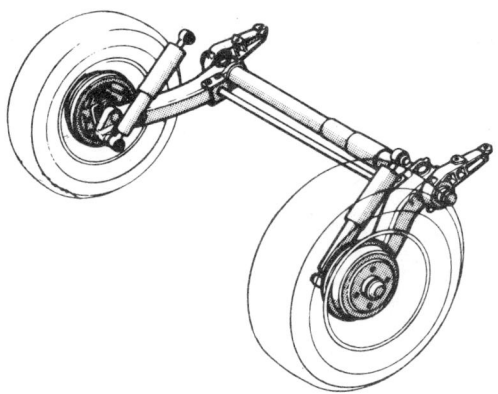

Fig. 33. Trailing arm layout for a Renault R5

Roll centre is at ground level, so large roll angle tendencies are usually resisted by some form of anti-roll bar. Several types of springing can be used with trailing arms, e.g. transverse, coaxial torsion bars (Renault R5), coil springs, Hydragas.

What is semi-trailing arm suspension?

This is the most popular form of IRS; it offers reduced camber changes and a compact layout. See Fig. 34. The pivot axis of the triangulated suspension arm is inclined to the centre line of the car; this raises the rear roll centre, whose height above ground level has several effects (see Chapter 2).

To accommodate the changes in drive-shaft length with suspension movements, a sliding, splined or plunge type joint is necessary in the drive shafts, or else the inner wishbone bearing has to be given freedom of movement by use of a swinging shackle (as in the early Ford Granada).

This type of suspension has inherent roll-stiffness, which gives improved road holding combined with low unsprung weight, thus providing a smoother ride. It is used by many manufacturers.

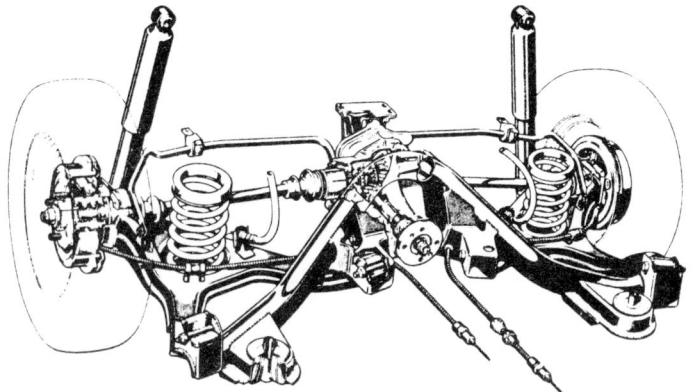

Fig. 34. *Semi-trailing arm independent rear suspension on Ford Sierra*

What is the unequal-length transverse links system?

Often referred to as the classic layout of IRS, this system has two transverse wishbones to locate the wheel upright (Fig. 35). Driving and braking torque is resisted by wide-spaced wishbone-mounting points. Generally used only on rather expensive sports and racing cars, its advantage over other layouts is the freedom it gives in choice of geometry to obtain the best compromise in road holding and ride comfort. A disadvantage is the extra space required for the upper arm.

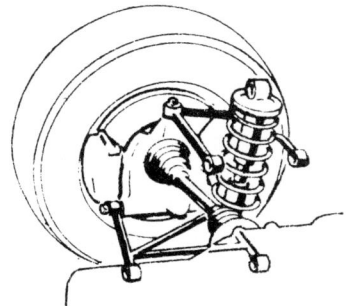

Fig. 35. *Unequal-length transverse links*

An alternative layout uses the drive shaft to serve as the upper link (Fig. 36). This saves space, but with the drive shafts absorbing side loads, stronger and more expensive differential side bearings are required. Note the use of a fixed-length drive shaft.

What is the transverse link and strut layout?

This is similar to that used at the front, except of course the strut cannot be steered. The use of this layout is more common on FWD cars, where wasted space above the wheels in the luggage compartment can be utilised by the strut and springs. Sometimes the springs are separated from the strut, as in the Ford Escort. See Fig. 37.

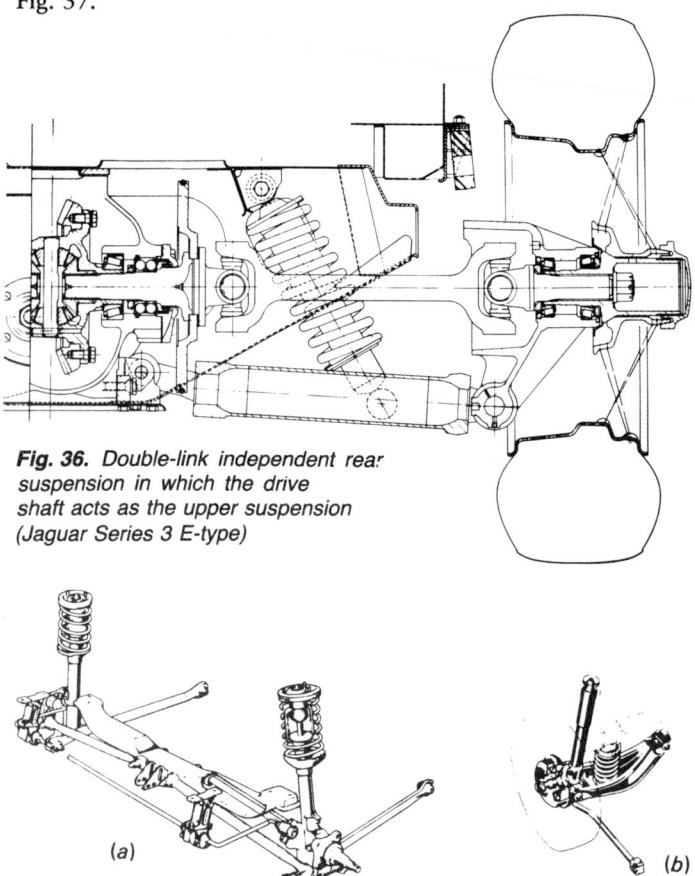

Fig. 36. *Double-link independent rear suspension in which the drive shaft acts as the upper suspension (Jaguar Series 3 E-type)*

(a)

(b)

Fig. 37. *Two types of transverse link and strut rear suspension: (a) Lancia Delta (b) Ford Escort*

Why has the swing axle fallen into disuse?

The swing axle has a very high roll centre, and a lot of camber change. Owing to the high roll centre it also 'jacks up' under hard cornering.

How does the swing-axle suspension jack up the car?

When the car is cornering a lateral force acts at the contact patch of the tyre, and this is in the opposite direction to the centrifugal force that makes the car roll. Because the centrifugal force acts above the hub level, the forces combine to tend to make the wheel lean outwards and lift up the inner pivot of the swing axle. As a result the car is jacked up (Fig. 38). As this happens the wheel tends to lean outwards with positive camber, and it can then dig in, causing the car to overturn. This happens rarely, but it is an inherent fault with swing axles.

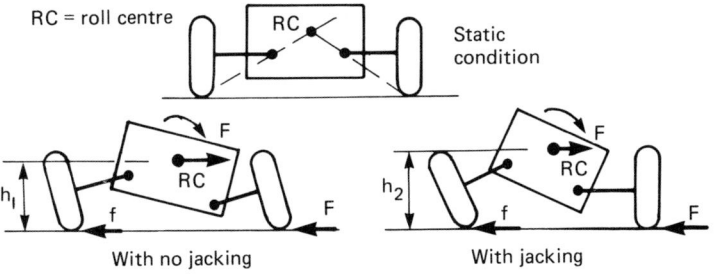

Fig. 38. *Jack-up with a swing axle*

Do other suspensions cause jacking up?

There is a certain amount of jacking up with most suspensions with roll centres above the ground, but it is only when the roll centre is at least as high as the hub that jacking up becomes serious.

Which geometry is best?

Suspensions that move the wheels up and down vertically are as bad as swing axles, in that on cornering the wheels lean outwards. For maximum cornering power the wheels should lean in slightly or remain at right angles to the ground. With a double-wishbone system it can be arranged that the wheels lean hardly at all on cornering, and this means that they will lean slightly on full bump and full rebound. The MacPherson strut suspension is not quite so good in this respect, but for practical purposes is not at a disadvantage. The geomery of the semi-trailing link is also good.

8
Dampers

What are dampers and why are they necessary?

Dampers, often referred to as shock absorbers, are devices that offer resistance to suspension movements, either hydraulically or by gas pressure. They are necessary to damp out the vertical oscillations of the suspension and car body after the springs have been deflected by road surface irregularities.

How much deflection takes place?

It depends upon the road surface condition, and the speed and mass of the car. When the spring is deflected it has a stored force

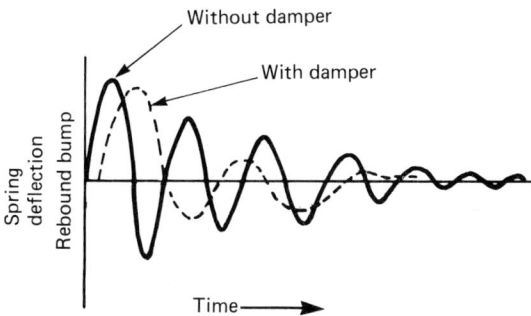

Fig. 39. *Effect of damper on spring oscillations*

equal to the spring rate and amount of deflection. For example, if the spring rate is 18 N/mm and the bump deflection is 25 mm the stored energy is 18 × 25 = 450 N. This energy then forces the roadwheel downwards with a rebound action, which not only further compresses the tyre but extends the spring beyond its normal static position, resulting in a further upward bump movement. This cycle of oscillations continues until the original energy is dissipated (Fig. 39).

What effects other than discomfort would be experienced if the damping action was insufficient?

Several serious problems affecting safety and tyre wear would be created. (1) The wheels may be lifted clear of the road surface during the continued oscillations; they cannot then steer or brake the vehicle. (2) Continued oscillations cause increased wear of all suspension components. (3) Increased body roll increases wheel camber and roll steering. (4) Excessive wheel movements cause increased tyre wear, uneven braking and poor acceleration. (5) There is increased pitching during acceleration and braking. (6) The large deflection of headlamp beams produces increased danger and aggravation during night driving.

How is the deflection energy dissipated?

By forcing hydraulic fluid at high velocities through small holes. This action generates heat, which is then lost to the surrounding structure and air flow.

Does the damping action remain constant at all speeds?

No, because the damper's resistance to deflection is a function of the square of its velocity. Thus little resistance is offered to relatively slow movements of wheels, but the resistance increases rapidly as wheel movement velocity increases.

What would be ideal damping?

For comfort, the damping should be light but stiff enough to prevent the body from floating. For good road holding and handling, the damping should be stiff but not so stiff as to interfere with suspension movements during cornering.

A compromise has to be made between these conflicting requirements. Some types of damper are adjustable for this reason, especially when the vehicle may operate in a loaded and unloaded state.

What types of damper are employed?

The original dampers, first introduced in the 1920s, employed friction. These took the form of alternate discs of steel and asbestos, interleaved to provide resistance to angular movement. Modern dampers are hydraulic.

Why are friction dampers no longer used?

A serious disadvantage with this type of damper is its high initial resistance to movement, which then reduces as movement continues. Static friction is higher than kinetic or sliding friction.

What form do hydraulic dampers take?

Two basic forms are used. The lever type can form part of the suspension linkage, i.e. the top wishbone. The telescopic type is a direct-acting dual or mono tube with or without gas assistance; it may also form the steering swivel (MacPherson strut).

Does the damper offer the same resistance on bump and rebound?

The damping on the bump stroke is often arranged to be less than that on the rebound stroke, because the bump stroke motion is frequently a violent forced motion compared with that of the rebound stroke, which is somewhat gentler.

What advantages does the telescopic damper have over the lever type?

An inherent problem with the lever-type damper is that it magnifies the effect of any slight reduction in damper resistance, owing to the lever arm arrangement. The telescopic type is connected immediately between the body and axle and is thus direct-acting. Several additional advantages are that it is more compact, and may be mounted concentrically with the suspension spring; it uses a lower working pressure, giving reduced stress and increased reliability; and it can be mounted vertically or diagonally to assist axle or suspension location.

How does the lever-type damper work?

A lever is connected to the suspension (or forms part of the suspension linkage) so that it articulates on a spindle in the housing. As it articulates, it operates a rocker arm at each end of which is a connecting rod extending downwards to a piston. There is a passage connecting the two cylinders in which the pistons slide, and this incorporates the damper valving (Fig. 40).

When the wheel is deflected on bump, the bump piston moves downwards, pressurising the fluid, so that the compression valve is lifted off its seat and the fluid passes through the valve to the other cylinder. The rebound valve does not operate. On the rebound stroke the movement of the rebound piston pressurises the fluid so that the rebound valve is pushed downwards off its seat, and the fluid can flow back to the compression cylinder. Thus the compression and rebound valves are quite independent.

Does the damper operate on every bump?

When the wheel passes over very small bumps it is impractical for the valves to be opened, since if they operated at these very low

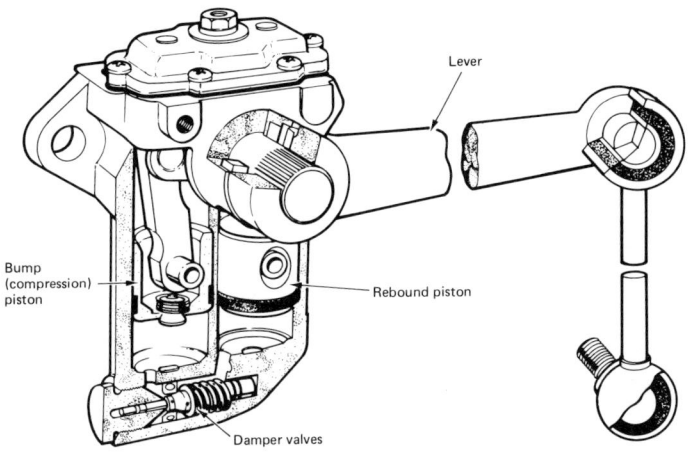

Fig. 40. *Armstrong lever-type damper, with two cylinders connected together by passages and the damper valve*

speeds and forces they would not be able to cope with the higher loads so well. Therefore a small notch allows some fluid to leak between the chambers when deflections are small.

How does the operation of the telescopic damper differ?

Although the principle of the telescopic damper is similar, the use of a *single* piston leads to a different arrangement. This piston is at the end of a rod that is normally secured to the car body, and it runs in a tube in a cylinder secured to the suspension arm. The annular space between the tube and the cylinder acts as a reservoir for the hydraulic fluid. Generally there are two sets of valves, one in the piston, the other in the base of the tube, but connected to the reservoir.

How does the dual-tube damper work?

When the wheel moves upwards on *bump* the damper is telescoped, which causes the piston to move downwards (Fig. 41).

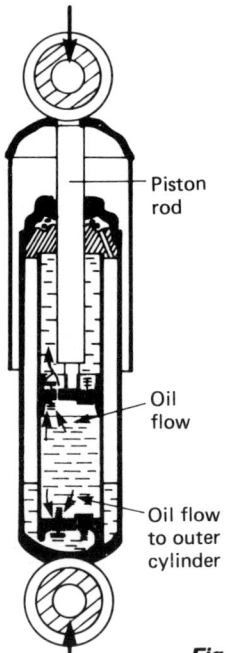

Fig. 41. *Damper operation on 'bump'*

Oil flows from below the piston upward through the orifice valves to the space above the piston, thus equalising the pressures. The piston rod also enters this space and displaces a quantity of oil to the outer cylinder or reservoir via the foot valve. The resistance offered by the foot valve to the oil flow gives the bump damping force.

When the wheel moves downwards on *rebound* the damper is extended, causing the piston to return upwards. This upward movement pressurises the oil above the piston and forces it to flow down to the space below the piston via the return orifice valves. The resistance offered to the oil flow gives the rebound damping force. The withdrawal of the piston rod from the upper cylinder space increases the cylinder's volume accordingly, and oil is drawn back from the reservoir via the foot valve.

Is the operation of a telescopic damper that simple?

In practice the actual design and operation are both very sophisticated, with the damper action being virtually tailor-made for particular applications. Damper settings are varied during the development stage of a new vehicle to obtain the best ride consistent with good road holding and cornering.

Do all telescopic dampers possess a dual-tube or reservoir arrangement?

No; many of the current designs employ only one tube, the internal construction of which allows them to be mounted at any angle, including upside down. The added volume of the piston rod is accommodated by the use of an inert gas, usually nitrogen, under compression at the bottom of the tube and separated from the oil by a floating piston.

How does the Monitube work?

The main operating difference between the Monitube (Fig. 42) and the conventional damper is that it has all its valves in the piston, and does not have a foot valve. In fact, the normal position for the Monitube is upside down, i.e. with the gas space beneath the working cylinder. The outer tube is also the working

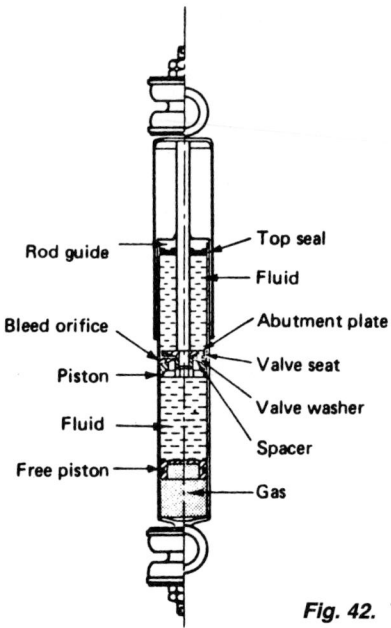

Rod guide

Bleed orifice

Piston

Fluid

Free piston

Top seal

Fluid

Abutment plate

Valve seat

Valve washer

Spacer

Gas

Fig. 42. *The Girling Monitube damper*

cylinder in which the piston runs, and the piston rod is much more robust than normal. The piston rod passes through a seal in the cylinder.

How is the gas separated from the fluid?

There is a thin, flexible diaphragm or a floating piston between the gas and the fluid, the gas being pressurised to about 1400–2100 kN/m². Thus it is impossible for the gas to mix with the fluid.

66

Are there any other advantages with the gas damper?

There are two other main advantages with this type of damper, which is made with minor differences by a number of manufacturers. First, since the outer tube serves as the working cylinder, much more fluid is in use in a given size of damper – generally about 40 per cent more. Secondly, the heat path between the working fluid and the air is more direct, so the damper runs at lower temperatures than normal dampers. The unsprung weight (i.e. the piston rod in the case of the Monitube, as against the rest of the damper on the conventional unit) can also be reduced.

Are there any disadvantages with these dampers?

The potential disadvantage of the gas damper is that the seals on the piston rod must be able to operate at higher pressures. This is a design problem, and in practice it means that, once there is a leak, performance will decrease very rapidly.

How can damper performance be checked?

Although the Shock Absorber Manufacturers' Association has devised a test and its associated equipment, an approximation of the damper's performance can be made by pushing the front (or rear) end of the car down a few times and then releasing. The car should immediately adopt its normal static position. If the car continues to bounce up and down the damper(s) may need replacing.

What are load levellers, or load-adjusters?

These are telescopic dampers that incorporate a coil spring, often progressively rated. The damper carries out its normal duties, while the auxiliary coil spring assists the main suspension springs.

What types of vehicle employ load-leveller dampers?

Any vehicle that operates in a heavily laden condition can have these dampers substituted for the normal variety. Load levellers are eminently suitable for cars towing a caravan, estate cars and lightweight vans and trucks. The ride will be a little harsher than normal, owing to the extra suspension stiffness, but the added safety, reduced wear and correct headlamp-beam alignment more than compensate for this.

9
Introduction to steering

What are the important features of the steering system?

It provides the driver with a means of maintaining directional control at all times, including the ability to change the vehicle's path, with a minimum of effort. It must also be self-centring.

How have steering systems evolved?

Horse-drawn vehicles employed a centrally pivoted axle, but this arrangement required the wheels to pass under the body and did not give support to the body except at the pivot point. With the advent of the self-propelled vehicle some means of steering the wheels from within the vehicle, and the provision of separate pivots for each wheel, became necessary.

What is the first requirement of a steering system that uses separate pivots for each wheel?

All the wheels on a vehicle must turn about one single centre, so the steered front wheels cannot remain parallel during a turn, and in fact generally toe-out slightly if true-rolling motion is to be obtained.

Figure 43(a) shows that if the two front wheels turned through the same angle their paths would cross. This is not possible physically, so the tyres would scrub across the road surface and

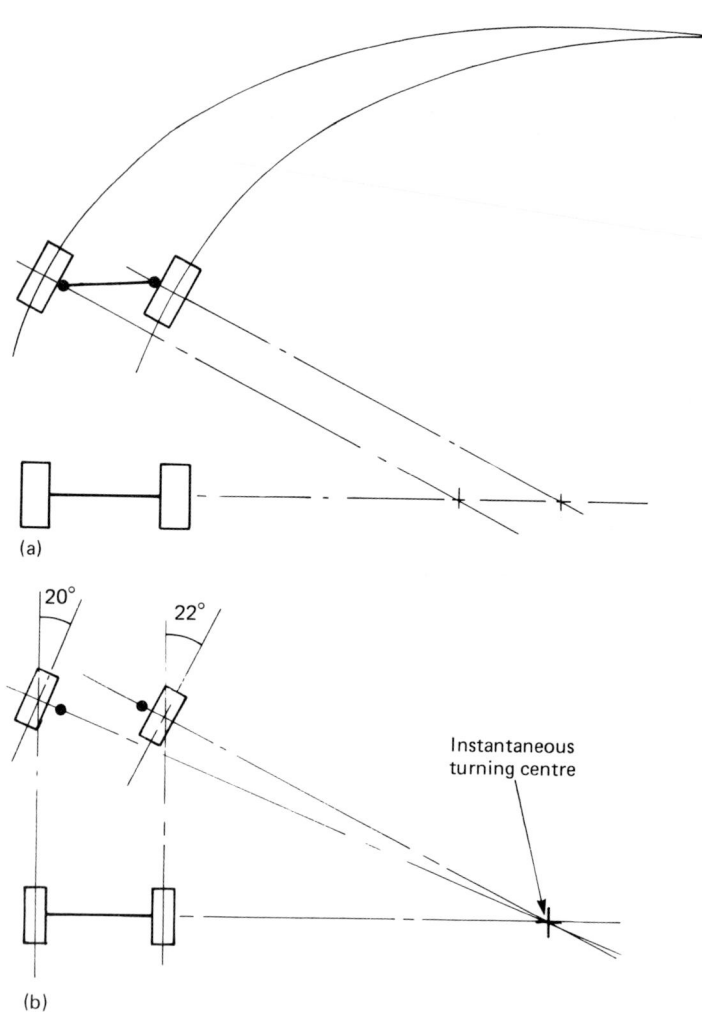

Fig. 43. *Turning (a) without and (b) with Ackermann angle*

cause increased tread wear. Figure 43(b) shows that if the inner wheel can be made to turn through a greater angle than the outer wheel, then all the wheels will turn about a common centre – the instantaneous turning centre.

How can the inner wheel turn through a greater angle than the outer when cornering?

The design of a linkage to carry out this requirement was patented in 1817 by Rudolf Ackermann and later improved in 1878 by Charles Jeantaud. In practice this arrangement gives exact angular wheel movements in one position on either steering lock.

The Ackermann principle requires the track arms to converge to a point near the centre of the rear axle (Fig. 44). The exact position of this intersection point depends upon the vehicle's wheelbase and track measurements and other factors.

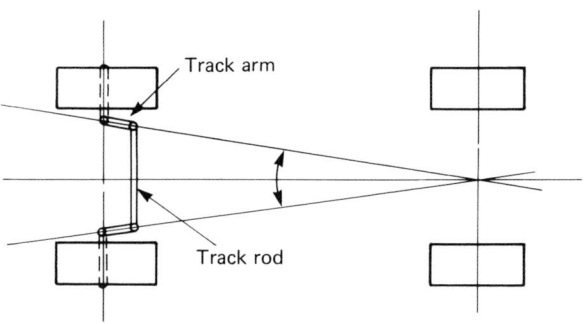

Fig. 44. *Ackermann principle*

Reference to Fig. 45 shows that when the track rod is moved a given distance to the left, the left-hand track arm moves towards its effective crank-angle (90° to the track rod) with a small angular movement, whereas the right-hand track arm moves

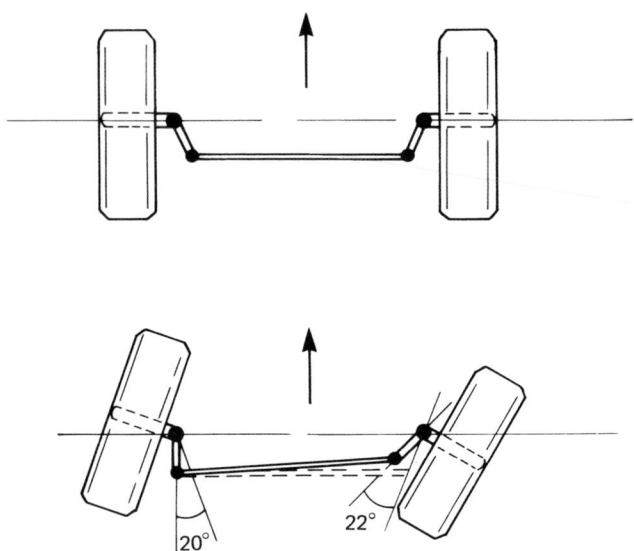

Fig. 45. *How the Ackermann principle produces toe-out on turns (T.O.O.T.)*

away from the effective crank angle and therefore increases its angular movement. It is this difference in angles moved by the track arms that turns the wheels by differing angles.

Note how the track rod does not remain parallel to the axle or suspension beam, but all the wheels align themselves at 90° to their radius of turn.

Do all vehicles employ the Ackermann principle?

Not any more, because three motoring conditions have changed: (1) vehicle cornering speeds are higher, so centrifugal force deflects the vehicle from a true line; (2) low pressure tyres are now used, which distort to resist side thrusts; (3) with relatively soft suspensions, wheel camber angle variations cause roll-steer effects.

How can the Ackermann angle be checked?

Place the front wheels on radius plates or turntables so that when the inner wheel is turned to 20° the outer wheels should have turned through about 22°. This check should be carried out on both left and right locks. They should be the same if the system is in good order.

How does rotation of the steering wheel actually turn the front wheels?

The steering wheel is fixed to a shaft or column which is part of a steering-box assembly consisting of some form of reduction gear. Rotation of this reduction unit creates partial rotary movement of an output shaft. This motion is then transmitted to the wheels by one of several means, depending upon whether a solid axle beam or independent suspension is used.

With a beam axle (Fig. 46) motion from the steering box is transmitted through a system of linkages – a drop arm, drag link,

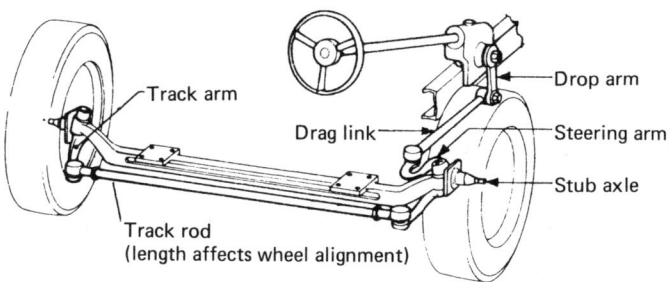

Fig. 46. *Beam axle steering*

track arms and track rod. The most common form of steering box used today is the rack and pinion (Fig. 47). Rotation of the reduction gear moves the toothed rack to one side, and this motion is transmitted direct to the track arms by short outer track rods.

See Chapter 10 for more detailed information on steering gears.

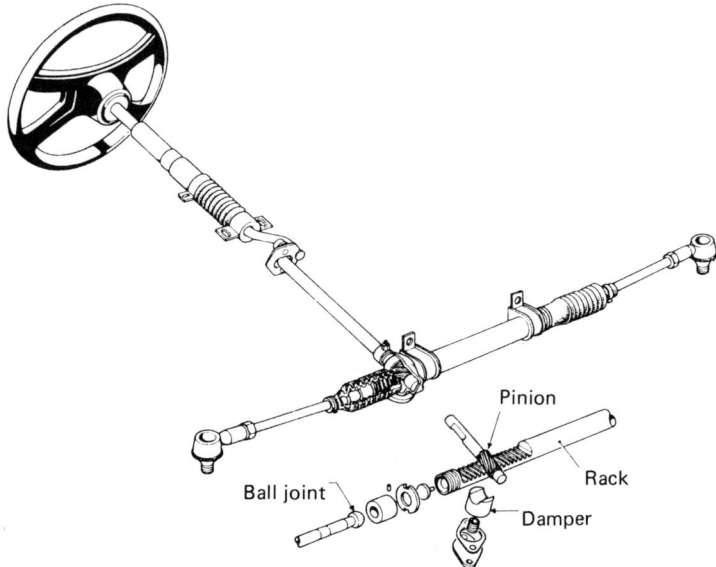

Fig. 47. *Rack and pinion steering*

Besides the Ackermann principle, what are the other elements of steering geometry?

For the steering to be precise, easy to handle and with a self-centring ability, the steering-pivot and wheel axes are invariably inclined to the vertical. The resulting features come under the headings of camber angle, king-pin inclination (KPI), offset, castor angle, and toe-in and toe-out.

What is camber angle?

Camber is the inclination of a road wheel away from the vertical when viewed from the front. If the wheel leans outward at the top it is referred to as positive camber (Fig. 48), negative camber if it leans inward. Camber angle has been gradually reduced from as much as 8° positive to something like 0–2° positive today. Too much camber increases tyre wear on the tyre's outer edge.

A cambered wheel tends to roll in the direction towards which it is leaning, thus producing a side force or 'camber thrust'. If the camber is positive, the wheels tend to turn outwards and keep all the steering joints slightly under load.

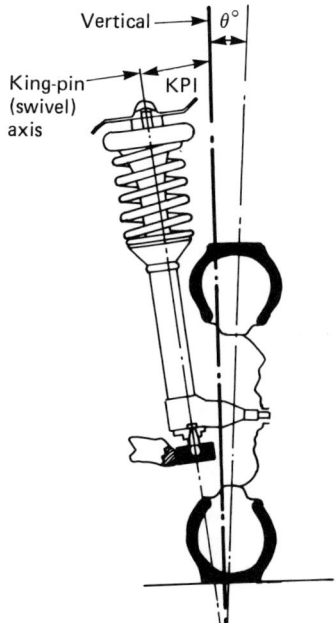

Fig. 48. *Positive camber angle (θ) and king-pin inclination*

Why is camber employed?

Camber was originally used to position the load more centrally over the tyre's contact patch, and to maintain the wheel perpendicular to the heavily cambered road surface. Because of the space occupied by the brake assembly and the steering pivots or swivels, it was not possible to position the load directly over the tyre's contact patch, and a large offset was produced between the swivel centre-line and the contact patch. This causes heavy steering and an increased tendency for the wheels to turn out.

Use of a dished wheel enables a smaller camber to be used, but the dish effect has to be somewhat limited so as not to seriously weaken the wheel structure.

Is positive camber always used?

Positive camber slightly reduces the cornering power of a tyre but makes it less sensitive to small side forces caused by road surface ridges and side winds.

Negative camber is often used to alter the handling characteristics of a car by increasing a tyre's cornering power and thereby influencing understeer or oversteer effects. The camber may even change from positive to negative or vice versa during suspension movements and body roll when cornering.

What is king-pin inclination?

This is the inclination of the king-pin or swivel axis relative to the vertical, when viewed from the front (Fig. 48). The need to accommodate the bulk of the brake assembly makes it necessary to incline the king-pin axis inward at the top so as to position the load near the tyre's contact-patch centre; this minimises steering effort.

Are any additional benefits gained by providing king-pin inclination?

Yes. With the king-pin or swivel axis inclined, the front wheels attempt to move downwards when they are turned or pivoted; as the road surface prevents this, the vehicle will be lifted. The vehicle's weight always resists this tendency and attempts to return the steering to the lowest and most stable position, i.e. to point the wheels straight ahead. Thus on either lock there is an inherent self-centring tendency.

What is centre-point steering?

When the centre-lines of both the swivel axis and cambered wheel coincide exactly at the road surface, the construction is said to give centre-point steering.

How can centre-point steering be arranged in practice?

There are three ways: (1) Keep the swivel axis vertical and incline the wheel outwards at the top (positive camber). (2) Incline both swivel axis and wheel to reduce positive camber, and introduce king-pin inclination. (3) Locate the swivel axis in the central plane of the wheel, using a deeply dished wheel.

The second method is the one generally used, but even then not all manufacturers are in agreement about the desirability of true centre-point steering.

What is meant by scrub radius or offset?

It is the distance between the swivel axis and the centre of the tyre contact patch (Fig. 49). Offset gives the steered wheels a rolling action around the swivel axis when they are turned.

Most vehicles employ positive offset, in which the swivel axis meets the ground inside the contact patch centre. Positive offset

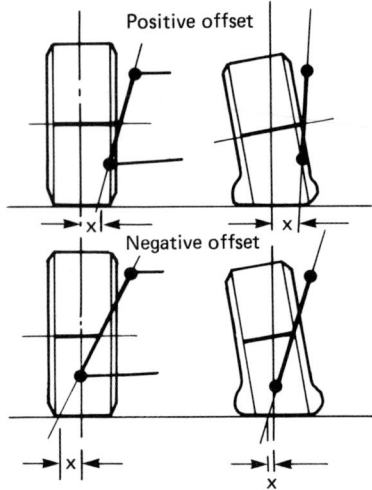

Fig. 49. *Positive and negative offset, and the effects of tyre deflation on offsets*

should not be too large, otherwise the steering will become heavy. The angle of inclination is usually about 8°, which gives an offset of about 30 mm.

Is negative offset used?

Sometimes referred to as 'over centre-point' steering, a negative offset is used by several manufacturers to obtain greater safety in the event of a tyre blow-out, where the sudden increase in a positive offset pulls the car violently to that side, with perhaps a loss of control. A negative offset, however, creates a reduced offset if a tyre blows out, so that less leverage is available for the deflated tyre to change the vehicle's direction. Figure 49 shows how the offset is changed by a tyre deflation.

Negative offset also gives better stability in the event of unbalanced braking, whether caused by road or braking-system conditions. Cars with dual braking systems often have a diagonal split, and improved straight-line braking is obtained if a failure in one part of the system occurs.

What is castor?

See Fig. 50. Castor is given to the steered wheels in an attempt to maintain the wheels in a plane that is parallel to the direction of motion. To enable the drive to 'feel' the stable straight-ahead position, a torque must be exerted on the steering wheel to overcome the self-centring or castoring action.

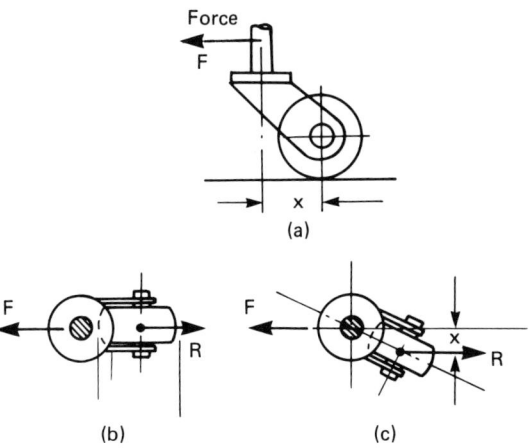

Fig. 50. *(a) The furniture castor illustrates the trailing action of the wheel as it follows the applied force, F. The distance x is referred as trail, and the castor is said to be positive. (b) The force R is aligned with force F when the wheel is in the straight-ahead position. (c) A self-aligning torque produced by the couple R and x tends to return the wheel back to the plane of motion*

What is castor angle?

Castor angle is the rearward inclination given to the swivel axis, so that the swivel centre-line coincides with the road surface in front of the tyre's contact patch (Fig. 51). The wheel will then follow the path taken by the swivel centre-line. The angle is measured at the top of the swivel axis, and is usually between 1° and 5°.

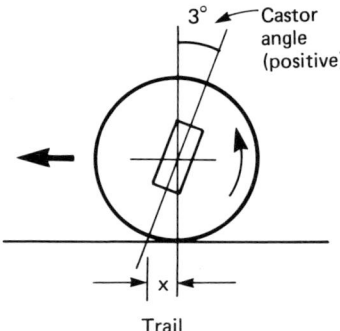

Trail

Fig. 51. *Positive castor*

How is castor arranged on a motor vehicle?

Although a similar arrangement to the furniture castor could be provided for steering a motor vehicle, it is simple and causes fewer operating problems to incline the king-pin or swivel axis in such a manner that its projected centre meets the road surface ahead of the wheel's vertical centre-line.

What are toe-in and toe-out?

When the vehicle is moving in a straight line both front wheels must remain parallel, but the use of positive camber combined with the tyres' rolling resistance results in the front wheels

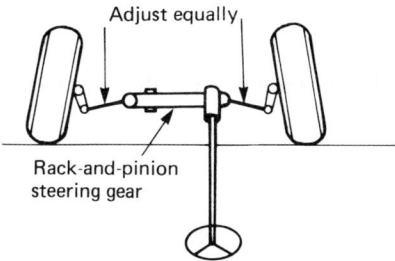

Rack-and-pinion
steering gear

Two-piece trackrod

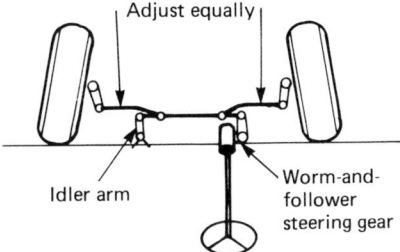

Idler arm

Worm-and-
follower
steering gear

Three-piece trackrod

***Fig. 52.** Adjustment of toe-in*

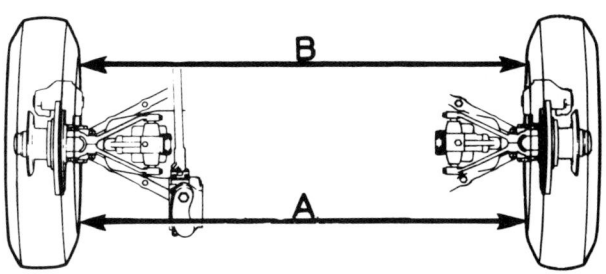

***Fig. 53.** Measurement of toe-in*

81

splaying out, or inwards if negative camber is used. This turning effect is due to the small clearances in each steering joint being taken up. This applies to rear-wheel drive vehicles; if front-wheel drive is used the wheels tend to turn inwards because the driven wheels' tractive effort acts in opposition to the tyres' rolling resistance.

How much toe-in or toe-out is generally employed and how is it measured?

On a RWD vehicle the front wheels normally move outwards about 2–3 mm, so a toe-in of 2–3 mm is given to compensate; a similar amount of toe-out is given on a FWD vehicle.

The toe-in or toe-out is measured at hub height on the rim of the wheel after wheel run-out has been checked. The actual adjustment is carried out by altering the lengths of the outer track rods by equal amounts (Fig. 52). Toe-in may also be defined as the difference between A and B in Fig. 53.

10
Steering gears

Have steering gears changed much over the years?

The advent of independent front suspension made a big diffference to the installation of the steering box, and to the linkage required. At the same time the continuing demand for precise but light steering also led to changes in the design of the steering box.

What mechanisms are used for the steering gears?

Many different types of mechanism have been used as steering gears, and as the name implies these tend to be based on gears of some type. Such systems are: worm and wheel, cam and roller (and cam and peg), screw and nut, recirculating ball, and rack and pinion.

What does the steering box do?

There are two basic functions of the steering box: to change rotary motion into linear motion, and to act as a reduction gear. If the steering ration were 1:1 (i.e. with a quarter turn of the steering wheel equalling a quarter turn of the road wheel about the swivel axis) the car would be very difficult to control. First, the amount of effort required, especially at low speeds, would be enormous, while at higher speeds the slightest movement of the steering wheel would send the car off the road. Therefore the steering box usually has a reduction ratio in the order of 15–25:1, so that you may require between two and five turns of the steering wheel to turn the road wheels through about 70°. In fact, only the rack and pinion steering gear converts rotary motion directly into linear motion, the other types relying on a linkage to do this.

How does the worm and wheel system work?

As its name implies the worm and wheel mechanism consists of a wormgear on an extension of the steering column, which meshes with a wormwheel or part of a wormwheel. On the shaft of the wormwheel is the drop arm that actuates a drag link to operate the steering linkage. This type of steering gear was used on many early cars, but has long since been superseded.

How does the cam and roller system work?

In many ways the cam and roller system is an improved worm and wheel. On the end of the steering column is an hour-glass-shaped cam in which a spiral groove is cut (this replaces the worm gear), and on the drop-arm shaft is a roller, which is V-shaped so that it engages with the spiral groove of the cam. Thus rotation of the steering column causes the roller to move up the spiral groove, so that the drop-arm shaft rotates. This system, which is incorporated in the Marles designs, has some advantages over the worm and wheel. Friction is kept to the minimum because the roller is free to rotate on its spindle, while the spiral groove can be formed to give linear or non-linear geometry. For example, it can be arranged so that the gear ratio reduces as more lock is applied. Therefore a few degrees movement at the straight ahead position would cause very little movement of the wheels, but near full lock the same movement would cause the wheels to turn more.

The cam and peg is similar, except that there is a peg on the end of a lever on the drop-arm shaft, and this engages with the cam.

How does the screw and nut system work?

In the screw and nut (Fig. 54) a screw thread is incorporated in the steering column, and it carries a nut. This nut cannot rotate, since it is attached by means of a pin and socket to a lever on the drop-arm shaft. Thus as the screw is turned it carries the nut along with it and the drop-arm shaft is rotated. One of the disadvantages of this design is that the large contact area between the screw and nut can lead to heavy steering.

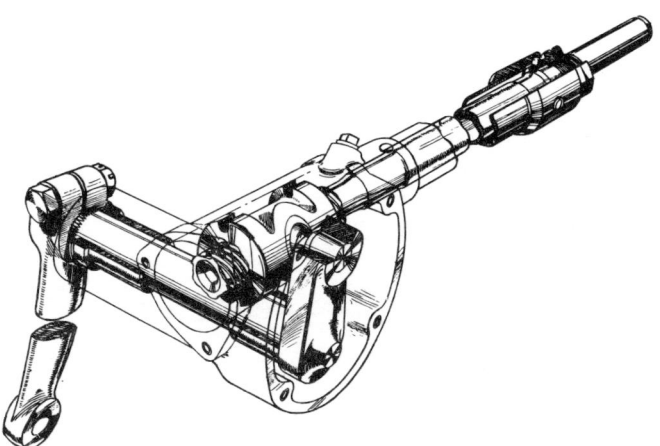

Fig. 54. Early Burman-Douglas screw and nut steering gear

How does the recirculating-ball system work?

The recirculating-ball steering mechanism (Fig. 55) has something in common with the screw and nut, and was designed to overcome the weakness of excessive effort. A screw or cam is incorporated in the steering column, and the nut is replaced by a

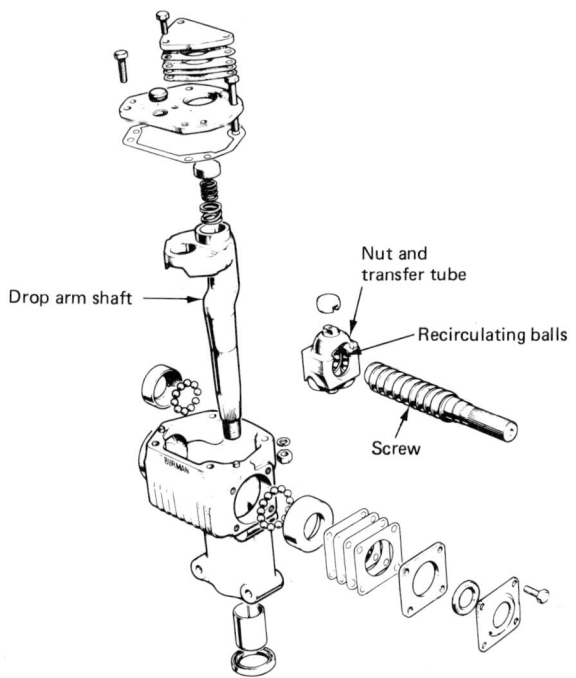

Drop arm shaft

Nut and
transfer tube

Recirculating balls

Screw

Fig. 55. *Exploded view of Burman recirculating-ball steering gear*

half-nut in which there is a groove, and a transfer tube which
closes the open side of the half-nut. In the groove are a number of
ball bearings. Since these balls are held in the groove and transfer
tube they act as a nut, and move the lever of the drop-arm shaft
along the screw as rotation of the screw causes them to roll
around the groove and transfer tube. The advantage of this design
is that the balls create very little friction, so the steering can be
very light.

How does the rack and pinion system work?

In this design (Fig. 56) a pinion is formed at the end of the steering column, and it meshes with teeth cut on a rack. With independent suspensions the rack is mounted transversely, and helical teeth are generally used. Rotation of the pinion causes the rack to move linearly, this being the most direct form of steering there is.

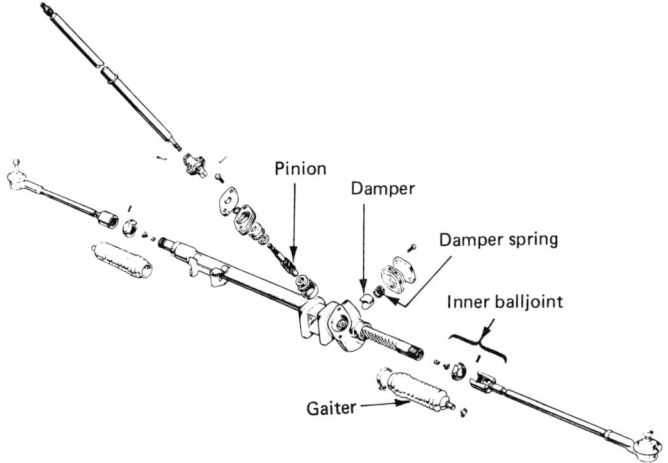

Fig. 56. *Exploded view of typical manual rack and pinion steering gear*

Which is the best form of steering?

Since it has the simplest linkage, with merely a track rod on each end of the rack, the rack and pinion is the simplest steering mechanism. For the same reason there is little lost motion or possibility of play or flexing in the mechanism, so it is mechanically sound. It can also be made so that it is self-compensating for the small amount of wear that does take place. On the other hand, the rack and pinion is fully reversible (i.e. it is just as easy to move the rack by pushing on the end as by rotating the pinion) and this means that more road shocks are transmitted through the steering than with some of the other systems. However this is a small disadvantage, and rack and pinion steering is now fitted to nearly all cars, including some large and expensive ones.

How does 'play' in the steering develop?

One of the major problems with the earlier types of steering gear was that 'play' used to develop quite quickly at the steering wheel – you could turn the wheel through, say, 15° before the road wheels started to turn. Play can be caused by a number of things: wear in the steering box, wear in the track-rod ends, or wear in the idler bushes, where an idler is fitted.

What components wear in the steering box?

The components most prone to wear in the steering box are the gears and the bushes. Roller or ball bearings can also wear, especially if they are overloaded. On modern gears the materials and treatments are such that wear is kept to very small amounts, even after high mileages, while the bearings are normally preset and so are also unlikely to wear.

How was wear compensated for on the old boxes?

In early designs the steering box could usually be adjusted to compensate for wear. The worm or screw was usually supported by a pair of ball bearings, and by adjustment of a screw it was possible to alter the amount of float of the screw on the shaft. With the worm and wheel it was possible to allow for gear wear by rotating the spindle for the worm, which would be eccentric to bring the teeth closer into mesh. On some of the cam and roller designs the roller was also carried on an eccentric spindle that could be rotated to take up clearance.

11
Power-assisted steering

What are the advantages of power-assisted steering?

There are three principal advantages: (1) a reduction of driver effort when steering heavy vehicles; (2) easier manoeuvring in confined spaces; (3) ability to use a higher steering-box ratio, which means less steering-wheel movement.

Are there any disadvantages?

Owing to the additional components required, the initial cost is higher and extra maintenance may well be necessary. Increased tyre wear may also be a problem, owing to ease of low-speed manoeuvring.

What are the basic requirements when designing a power-assisted steering system?

The driver must be provided with a feedback of information on road surface conditions (there must be a 'feel' to the system). The vehicle must be capable of being manually steered in the event of power-assistance failure.

Why is power-assisted steering and not power steering used on cars and trucks?

True power steering is a system in which there is no mechanical connection between the steering wheel and the road wheels. The driver merely opens and closes valves as he operates the steering, and fluid is then transferred to the axle to steer the wheels. In the event of a failure there is no steering left at all, so this type of steering is used only on slow-moving crawlers, earth movers and so on. With a power-assisted system there is a normal steering linkage, but some form of power is applied to the system so that the driver has to use very little effort to operate the steering gear.

What sort of power is used?

Normally a hydraulic power system is applied to the steering; on some commercial vehicles compressed air is used, although this is rare. In either case, there is a pump or compressor to pressurise the fluid to provide the necessary energy. The fluid is then supplied to a valve. In the straight ahead position the fluid is returned to the reservoir, but once the driver applies some effort the fluid is delivered to one end of a cylinder. The pressure acts on a piston and moves it along and in so doing actuates the steering.

How is a power-assisted steering system arranged?

It largely depends on the type of system used. It may be ram-operated linkage (Fig. 57) or integral within the steering box or rack assembly (Fig. 58).

In the Adwest power-assisted rack-and-pinion steering system the hydraulic assisting pressure is controlled and directed by a rotary valve. This valve (Fig. 59) is manipulated by the rotation of the steering column and applies the hydraulic pressure to one side or the other of a piston mounted on the rack unit. The rack unit is moved left or right by this hydraulic pressure in relation to the force applied by the driver to the steering wheel.

When the rotor is at rest (Fig. 59(a)), hydraulic fluid flows through ports A into the long grooves B and back to the reservoir without affecting the steering. When the rotor is turned (b), the fluid is diverted into the smaller ports C and is pressurised to move the rack of the steering mechanism. Fluid returns to the reservoir through ports D.

The steering column is connected to the valve through a torsion bar so that the assistance provided varies with the resistance of the tyres to steering.

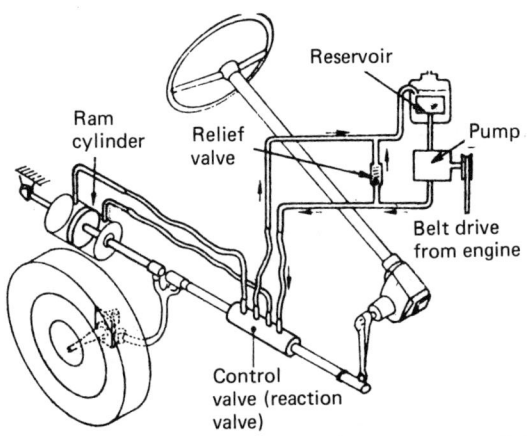

Fig. 57. *Power-assisted steering for a ram-operated linkage system*

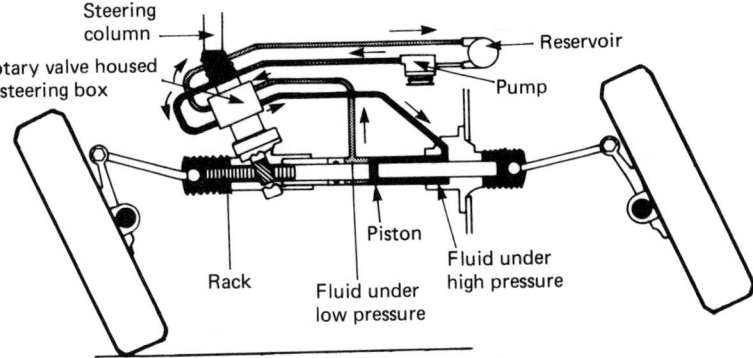

Fig. 58. *The Adwest power-assisted system*

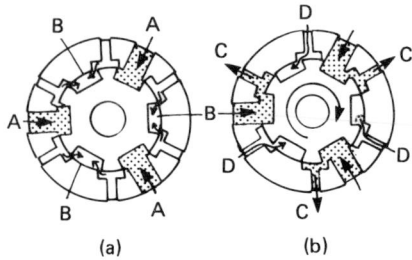

Fig. 59. *Adwest valve at rest (a) and when turned (b)*

Is the pump a separate self-contained unit?

On a car the pump is normally belt-driven from the crankshaft, and consists of a pump and reservoir. The pumps tend to have roller elements or rotors and vanes, and can deliver fluid at about 7000 kN/m². The pump delivers fluid through a high-pressure hose to the power-assistance cylinder or ram, and there is a return hose from the cylinder to the reservoir. See Fig. 60.

Where is the power-assistance cylinder mounted?

Power assistance can be applied in a variety of ways. In the early systems there was a separate hydraulic ram which was mounted separately, acting on the steering linkage. Then the ram took the place of a transverse rod in the steering. In more recent designs the cylinder is incorporated in the steering gear, be this recirculating-ball, cam and peg or rack and pinion. The control valve is also built into the steering gear, being installed so that it is operated when the steering column is rotated. In the recirculating-ball steering gear (Fig. 61) the fluid acts on a piston formed on the nut and thereby moves the nut axially, whereas manual effort rotates the screw to move the nut axially. In the rack and pinion gear the effort is applied directly to move the rack axially.

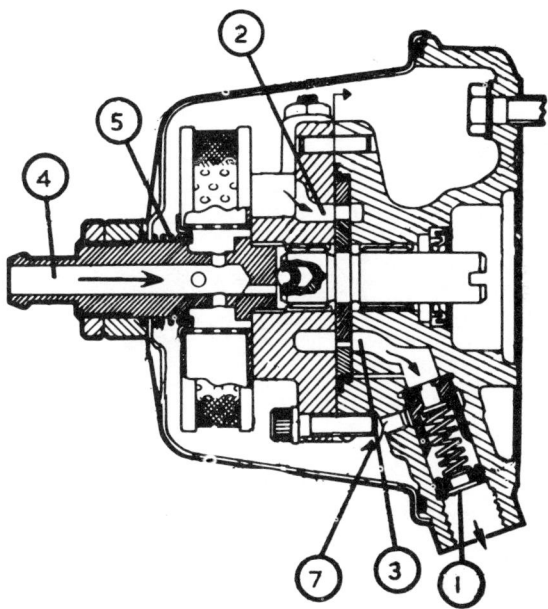

Fig. 60. *Power-steering pump/reservoir: 1 regulating valve; 2 inlet port; 3 outlet port from pumping element; 4 return from steering-gear power cylinder; 5 filter pressure-relief valve; 7 return from pressure-regulating valve*

How does the valve work?

The aim with the valve is to provide some 'feel' to the steering so that the more effort the driver applies, the quicker the system reacts. In the worst situation, operation of the valve would be like turning on a tap: full pressure would be applied however much the steering wheel was turned. The object with power-assisted steering, of course, is to reduce the steering effort, so there has to be a compromise between 'feel' and effort. The general way of

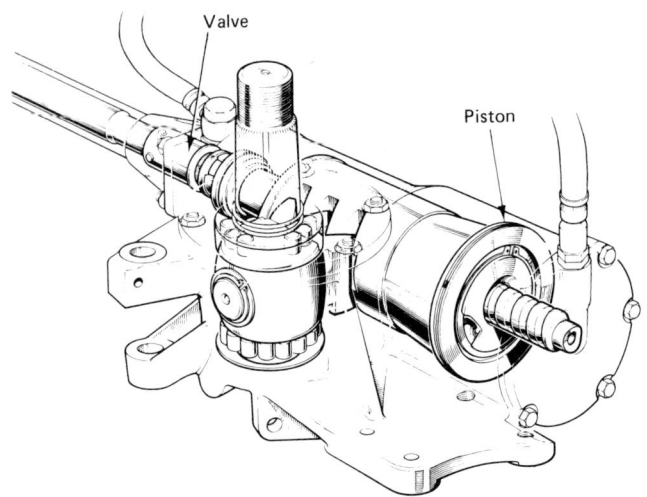

Fig. 61. *Integral power-assisted recirculating-ball steering gear (Burman)*

providing the feel on modern steering systems is for the driver to turn a slim torsion bar on the end of the column when he turns the steering wheel. This torsion bar forms part of the steering shaft, so when the driver twists it the friction between the tyres and the road surface provides some resistance. Only when the bar has been twisted by a certain amount will the valve be opened and will pressure be applied to the piston.

The effort required from the driver depends on the design of the torsion bar and valve: the stiffer the torsion bar, the more effort is needed.

What form does the valve take?

Although the valve is formed integrally with the steering gear, it usually has a separate housing bolted to the main housing. Projecting from the valve housing is a splined shaft, which is connected to the end of the steering column. This splined shaft is the input shaft to the valve, and pinned in its hollow end is one end of the torsion bar. The other end of the torsion bar is pinned to the pinion shaft in a rack and pinion unit, or to the steering screw/shaft in a steering box. When the torsion bar is twisted the input shaft rotates slightly relative to the pinion shaft (or steering screw). Fluid is delivered under pressure to an annular space in the housing, and it then passes through the valve to the power cylinder or to the reservoir.

How does the valve operate?

There are various designs of valve. In the Burman design (Fig. 62), for example, a portion of the valve is square-section and extends into the hollow pinion shaft. The fluid can enter the space between the pinion shaft and the square-section shaft, and can then pass out through other ports to the power cylinder. In the straight-ahead position the fluid is delivered through the inlet passage to an annular passage surrounding the pinion shaft, and then passes through a pair of diametrically opposed ports into spaces between the flat portions of the input shaft and the bore of the pinion shaft. Since the spaces filled with fluid are opposite one another there is no out-of-balance force; the fluid does not pass through the outlet passages to the power cylinder, therefore, but returns to the reservoir.

What happens when the driver turns the steering wheel?

As soon as the driver applies sufficient force to the steering wheel to twist the torsion bar, the square-section shaft rotates relative to the pinion shaft. The fluid can still enter through the same passages, but there is now a direct passage between one of the spaces and an outlet passage. Therefore some fluid passes under pressure through that passage to one end of the power cylinder. At the same time the other passage now connects with the low-pressure chamber between the other input shaft flat and the pinion shaft; fluid discharge from the power cylinder flows into

95

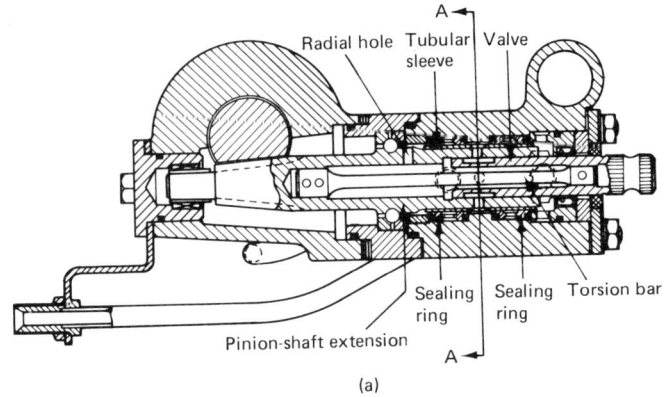

(a)

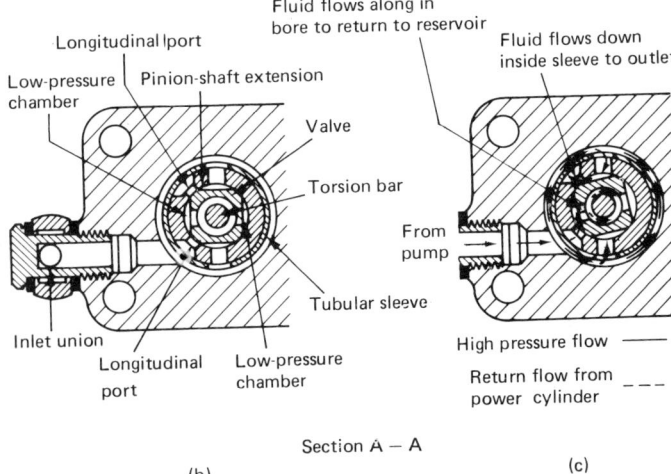

Section A — A

(b)   (c)

Fig. 62. *Power-assisted rack and pinion steering valve: (a) longitudinal section, (b) cross section in the straight-ahead position, (c) cross section when steering effort is applied (Burman)*

this space, and thence through radial ports in the input shaft to an outlet to the reservoir. If the steering wheel is rotated in the opposite direction, fluid passes to the other end of the power cylinder, while the fluid returns through the low-pressure chamber and out to the reservoir. In the Adwest design the principle is similar, but the internal shaft, instead of being square, has a number of grooves cut in it axially, and there are also some grooves in the bore of the pinion shaft.

How is the fluid contained in the valve?

To keep the fluid in the valve, and to prevent it leaking between sections of the valve, a number of high pressure seals are incorporated.

What happens if the torsion bar breaks?

Normally there is a safety device to ensure that the vehicle can be steered even if the torsion bar breaks. For example, on the Adwest design the end of the input shaft is splined to the pinion shaft, but there is sufficient angular or rotary clearance between the splines to allow the valve to operate; if the torsion bar were to break, the input shaft would take up the clearance and rotate the pinion shaft directly.

How is the power cylinder arranged?

In the integral recirculating-ball steering gear there is a piston on the nut; the piston runs in a cylinder that also forms part of the housing for the steering box. When the valve is rotated fluid flows under pressure to one end of the cylinder or the other, pushing the piston along and so forcing the fluid at the other end of the cylinder back to the reservoir. In the rack-and-pinion unit a disc on the rack usually serves as the piston, and again the pressurised fluid is delivered to one end or the other.

12
Stub axles and wheel bearings

What does the suspension linkage carry?

Whatever form of front suspension is used it must carry a member that in turn carries the wheel bearings, hub, wheel, and brake caliper or backplate. This member, or part of it, must be able to rotate on an almost vertical axis so that the wheels can be steered. On early independent suspensions the components are similar to those of beam axles, so the wishbones carry a vertical post, which pivots on bushes on the wishbones. There is a lug halfway up the post, which resembles the end of an axle. The king-pin or swivel-pin is anchored in this lug, and it carries the stub axle, which therefore rotates independently of the vertical post for steering. The stub axle incorporates a flange to carry the brake, and a spindle to carry the wheel bearings and hub. These components from the stub axle outwards are found on all front suspensions in one form or another.

How do modern suspensions vary as far as the stub axle is concerned?

When considering the stub axles of modern suspensions we need to consider double-wishbone and MacPherson-strut systems separately. In the modern double-wishbone system, ball-joints are used between the wishbones and the stub axle carrier/vertical post. These joints allow the stub axle carrier to rotate for steering, and also accommodate the suspension movement. In some cases

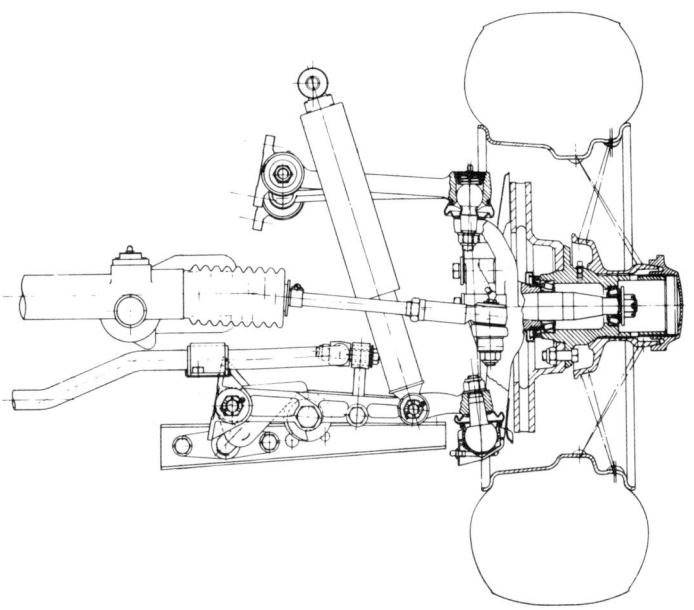

Fig. 63. *Jaguar Series 3 E-type front suspension, showing the wheel bearings and ball joints*

the ball-and-socket is built into the end of the wishbone, in others it is built into the stub axle carrier. The stub axle itself is either forged integrally with the steel carrier or is a spindle pressed into the carrier. With front wheel drive, the stub shaft must rotate to transmit drive, so the bearings are carried in a bore of the carrier, not on a spindle as on an undriven axle.

What is the arrangement with MacPherson-strut suspension?

With MacPherson-strut suspension the stub axle is attached to the bottom of the tubular strut – it is either welded or bolted on.

The bottom joint is a ball-joint, mounted either on the transverse link or on the stub axle itself. As with double wishbones, there is a steering arm that is used to steer the stub axle, and this may be integral with the stub axle or may be bolted on.

What sort of joint is used at the top of the strut?

The joint at the top of the strut must allow rotation of the strut for steering, must accommodate some angular movement and must carry the force of the spring. The joint itself consists of a rubber mounting, bonded to a flanged steel sleeve, and a thrust bearing. The flanged sleeve is bolted to the body. The piston rod of the strut terminates in a shoulder, a short portion of reduced diameter, and a threaded portion. This top portion of the strut passes through the thrust bearing, an abutment plate that carries the suspension spring being trapped between the shoulder and the thrust bearing. Thus the thrust bearing allows rotation of the strut, while the rubber mounting accommodates angular movement. Generally a ball bearing is used at the upper mounting (Fig. 64), but in some designs a thrust washer with low-friction facings is used, and on the Ford Escort and Capri there is no bearing at all, since specially shaped rubber mountings accommodate rotary movement as well as angular deflections (Fig. 65).

Are ball-joints used on all modern stub axles?

Although ball-joints are used for the upper joints of virtually all double-wishbone suspensions and in all MacPherson-strut systems, a different arrangement is used at the lower joints on some cars. The reason for this is that in most designs the lower joint has to carry the spring loads, and some designers prefer to apply these high loads to a ball-joint. On the Morris Marina, for example, a trunnion is used. The stub axle carrier terminates at its lower end in a threaded shank, and this engages in the threaded trunnion housing. There is a pair of lead bronze bushes mounted horizontally in the trunnion housing, and a plain pin bearing in these bushes serves as the pivot for the wishbone.

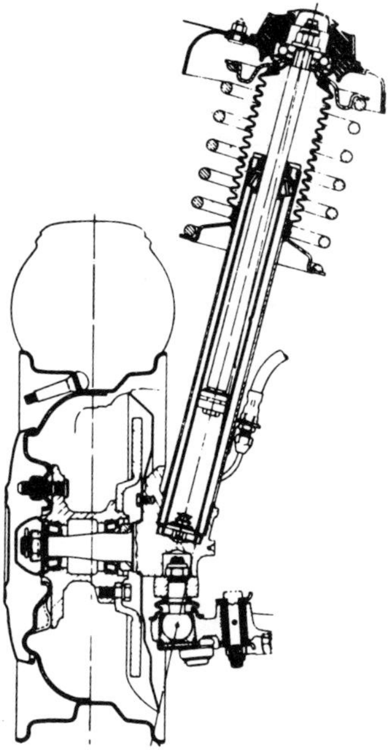

Fig. 64. Suspension strut with conventional bearing arrangement

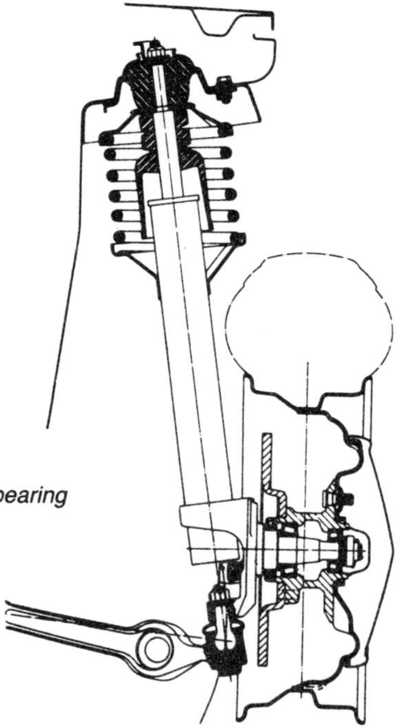

Fig. 65. Suspension strut with rubber upper joint without any bearing (Ford Capri)

101

What form does the ball-joint take?

Ball-joints vary significantly in detail, but the ball is usually formed at the end of a stud with a tapered shank (although again the Morris Marina is an exception, in that the shank is parallel); see Fig. 66. The tapered shank locates in a tapered bore either on the link or stub axle. The ball itself seats in a spherical housing; a spring-loaded pad holds the ball in contact with the housing, the pad also having a hemispherical recess to support the ball. In some cases the ball is in two parts, one half on the ball-stud, the other half in the form of a detachable bush.

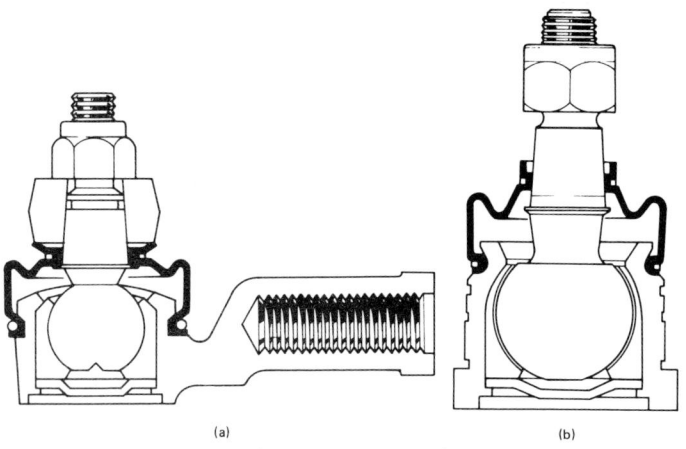

(a)   (b)

Fig. 66. Sectional ball-joints: (a) tie-rod end, (b) suspension joint

What sort of bearing materials are used?

Generally the ball-joints are sealed for life, and the ball-stud is invariably steel, but if it carries a bush this is likely to be of sintered iron with a slightly porous surface to retain lubricant. If the ball-joint is heavily loaded the cup in which the ball seats will almost certainly be hardened steel, but if it is lightly loaded a plastics cup may be used. An important feature of the joint is the rubber or plastics gaiter that prevents water or dirt getting into the joint.

What sort of bearings are used in the hub?

In rear-wheel-drive cars tapered roller bearings are almost always used to carry the hub and wheel, although angular contact ball bearings have been used on many cars.

Tapered roller bearings have a large area, and the effective span of the bearings is greater than with ball bearings. (The span is the distance between the points where lines drawn at right angles to the inclined bearing surfaces strike the axis of the spindle.) The hub is usually designed so that the inner bearing carries more load than the outer bearing, and for this reason the inner bearing is of larger diameter.

What form does the hub take?

The hub is usually a malleable iron casting, and is basically cylindrical with a flange for the wheel studs and another flange for the brake disc. The outer races of the tapered roller bearings are pressed into recesses in the hub, and there is a seal at the inner end of the hub. Since the inner races are a light push fit on the spindle, the hub can be built up as an assembly before being installed on the car. The bearings are retained by a nut, and a cap prevents the ingress of dirt and water. To prevent it from rotating the nut is retained by a split pin.

How is the wheel retained to the hub?

A special type of stud is used to secure the wheel to the hub. It has a small head and a partially serrated shank. It is pressed into the hub flange, the serrations preventing it from turning while the nut is tightened.

How is the wheel located?

The wheel nuts have tapered noses that bear in tapered holes in the wheel, and this gives a positive location. However, it has been found that this method does not give the accuracy needed with radial ply tyres, so in many cases the hole in the centre of the wheel is machined to locate on the hub.

What form do hubs take on front-wheel-drive cars?

On FWD cars the rotating portion of the hub consists only of the stub of the drive shaft and the wheel mounting flange, which is splined to the stub shaft. In some cases the hub of the wheel flange extends through the bearing and abuts against a shoulder on the shaft. In others the flange abuts against the bearing. The bearings are carried in the bore of the stub axle carrier. On the Austin Maxi a separate housing carries the bearings. On the Ford Fiesta a tapered roller developed by Timken is used. Called 'Set Right', the inner races of the bearings are extended inwards so that they abut against one another. The bearings are produced to very close tolerances and need no adjustment; they are locked up solid, the clearance being built into the bearings.

What bearings are used at the front in front-wheel-drive cars?

Angular contact bearings are usually fitted, because assembly is simple. On the Renault 12, for example, there are two separate bearings pressed into recesses in the stub axle carrier. There is also a spacer between the inner races of the two bearings, to make sure that there is no axial play, so that both bearings withstand lateral forces. If tapered roller bearings are used in a similar installation, shims are needed to get the correct preload or clearance, as the case may be, and this is a tricky operation.

Are there always two separate bearings?

The trend with wheel bearings on FWD cars is for the bearings to be supplied as a unit. On both the Fiat 128 and Austin Maxi there is a single outer race for the two bearings, and this is pressed into the stub axle carrier. There are two inner races, which extend inwards far enough to abut against one another. The balls sit in a groove in the inner race but, since the outer sleeve or race has only half grooves, the inner races, balls and cages can be installed separately. In the Fiat, seals are incorporated in the bearings, while in the Maxi two separate lip seals are used.

What types of bearings are used at the rear with FWD?

Hub arrangements similar to those at the front of rear-drive cars tend to be used. However, since the loads are generally much lower, a pair of angular contact bearings, or an angular contact double bearing, or small tapered roller bearings may be used.

13
Wheel and tyres

What are the basic requirements for a road wheel?

Besides withstanding, in conjunction with the tyre, the vehicle's weight, the wheel must be strong enough to transmit driving and braking torque and also resist side-thrusts during cornering. With these requirements in mind the final product should also be light in weight (to reduce unsprung weight), easy to clean and cheap to manufacture.

How many types of wheel are available?

There are three main types generally used for cars (Fig. 67). The most common is the pressed steel disc. Less common and very expensive is the wire-spoke. A cast alloy casting is popular for sports and racing cars, owing to its lightness.

Do commercial vehicles employ any of these car-type wheels?

Although lighter commercial vehicles use the pressed-steel disc, the fitting and removal of the stiff side-walled tyres used by the heavier vehicle requires the use of wheels with detachable rims which are locked in place by a flange or by a separate split locking ring – a two-piece rim or three-piece rim respectively (Fig. 68).

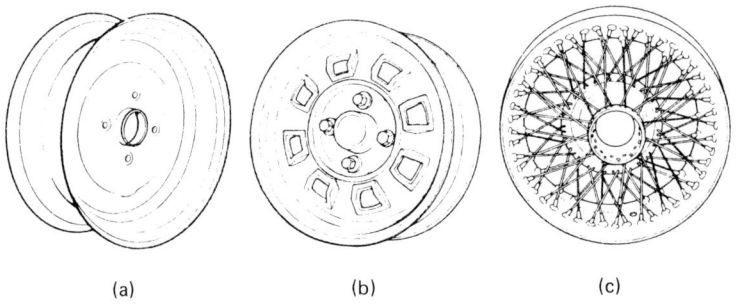

Fig. 67. Three wheel types for cars: (a) pressed steel disc, (b) light alloy, (c) wire spoke

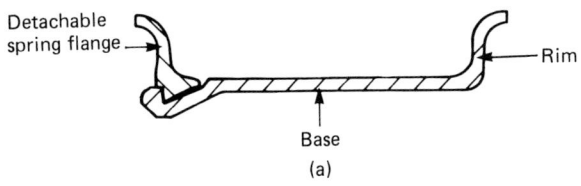

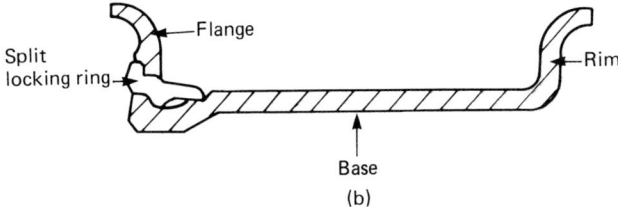

Fig. 68. Commercial vehicle wheel rims: (a) two-piece, (b) three-piece rim assembly

What are the essential features of a pressed-steel disc wheel?

Basically the wheel is dished to give increased strength, with a rolled section rim welded in place. A number of slots or holes are arranged under the rim to allow ventilation of the brake assemblies. A 'well' is included (Fig. 69) to allow the fitting and removal of the tyre whose rim cannot be stretched. The well accommodates one tyre rim, while the other is levered over the flange.

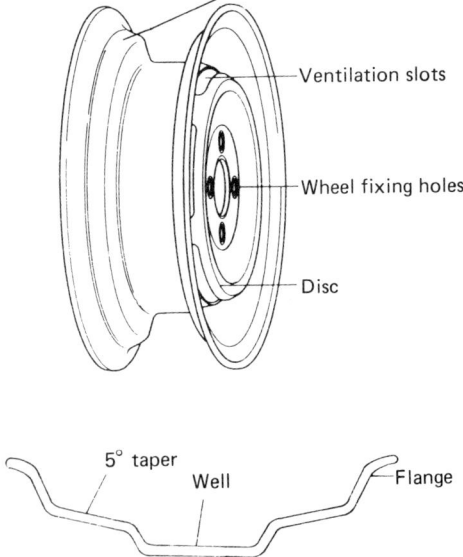

Fig. 69. *Typical well-base steel wheel*

The wheel rim is given a 5° taper to provide a good airtight seal at the bead of the tyre. In effect, the bead becomes firmly wedged on the rim as it moves up the taper during inflation.

How much weight is saved by cast alloy wheels?

This type of wheel can be manufactured from either aluminium alloy or magnesium alloy. Aluminium alloy gives a weight saving of about 30 per cent, and may be stove-enamelled to improve appearance and corrosion resistance. Magnesium alloy gives the even greater weight-saving of about 50 per cent when compared with a pressed-steel wheel.

The savings in weight mean that section thicknesses can be greater to give increased wheel rigidity. In a similar manner the rims can be increased in width to allow the use of wider tyres for improved road holding. It is the reduced unsprung weight that gives even greater advantages by allowing the wheel to follow the road surface irregularities more closely with less disturbance of the vehicle and its load.

Do cast alloy wheels have any disadvantages?

Besides being very expensive, cast alloy wheels have low resistance to corrosion and accidental damage. Magnesium is inflammable and easily damaged by salt.

With so many varieties of wheel diameters, widths and appearances, how can the correct wheel be identified?

A set of standard dimensions listing rim widths and associated tyre sections has been drawn up by the SMMT (Society of Motor Manufacturers and Traders). Part of the SMMT standards table is shown below.

Rim code*	Rim width	Flange height	Well depth	Rim diameter
4½J–13	4.5 in	0.687 in	0.75 in	13 in
5J–13	5.0 in	0.687 in	0.70 in	13 in
5K–13	5.0 in	0.77 in	1.00 in	13 in
6L–15	6.0 in	0.85 in	1.125 in	15 in

* First number is rim width in inches; letter indicates rim flange height; second number is wheel rim diameter in inches.

Has there been any further improvement in rim design?

In recent years, alterations have been made to make it more difficult for the tyre to be forced away from its rim during fast cornering. Figure 70 shows a range of safety rims for cars. The long flat ledge prevents the tyre being forced into the well by locally applied force arising from cornering. The bead can only be forced into the well if the entire bead is pressed at the same time along the flat ledge and into the well. Some rim designs incorporate 'humps' to prevent tyre movement away from the rim.

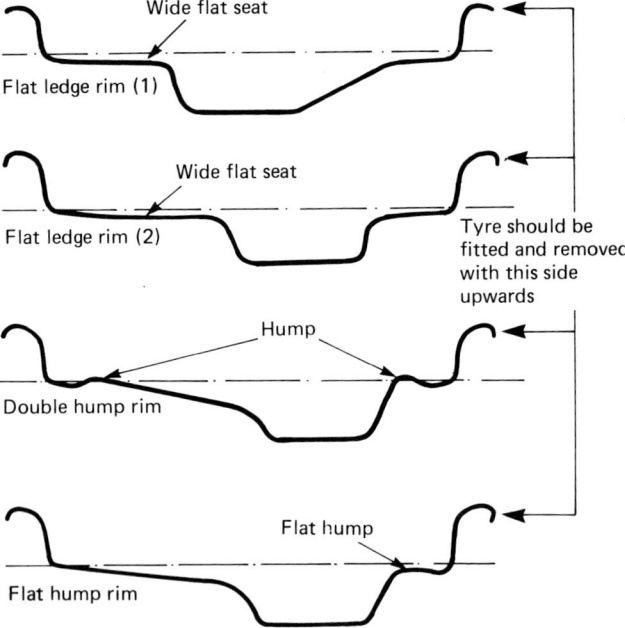

Fig. 70. Safety rims (Dunlop Ltd)

Are there any further developments in safety?

When the tyre suffers the type of puncture that results in sudden deflation while still in motion, the possible risk of the beads dropping into the well and then rolling off could be a problem. Recent rim developments have concentrated on designing a wheel rim shape which prevents the tyre leaving the rim in these circumstances.

Such a rim has been jointly developed by Dunlop and Michelin and called the TD wheel, which can only be fitted with a TDX tyre – a tyre-wheel combination. The TD wheel allows the driver to continue his journey at a reduced speed, i.e. up to 60 km/h (40 mile/h), for some distance before changing a deflated assembly at a convenient stopping place.

Is it really necessary to know the difference between one type of tyre and another?

From the safety viewpoint it is important to know the differences between types of tyre because they have very different road holding, steering and braking performance. For instance, if an incorrect mix of tyre types is made on a vehicle, this could well result in complete loss of control when cornering.

How many types of tyre are available?

There are three main ones (including one used primarily in the USA). *Cross-ply* tyres were first used in about 1920, using rubber-coated textile cords to replace the canvas fabric then used. Michelin introduced the first low-pressure tyre in 1923. In 1948 Michelin introduced *radial-ply* tyres; these generally use rubber-coated rayon textile cords running from bead to bead, i.e. at 90° to the circumference, with a layer of tread bracing cords running round the circumference.

Bias-belted tyres are a compromise between cross-ply and radial-ply, manufactured on existing equipment in America. The

111

plies run diagonally, with a tread-bracing layer running round the circumference.

What are standard terms used when describing the technical features of a tyre?

The following terms are illustrated in Fig. 71.

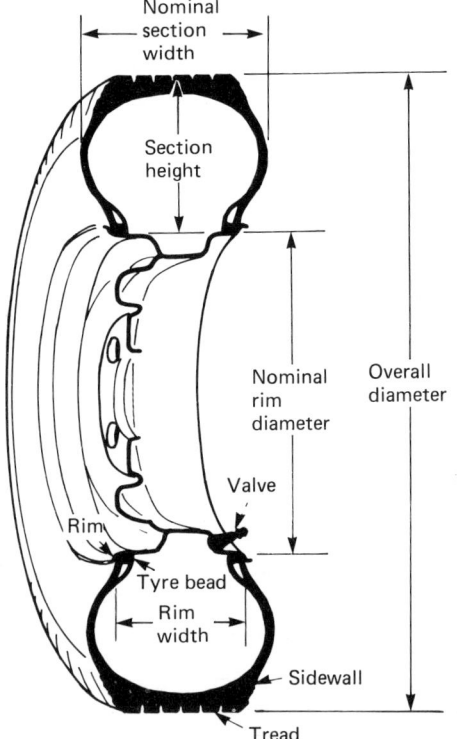

Fig. 71. *Standard terms for describing tyres*

(1) *Nominal tyre section*. The width of a tyre nominally stated, because the actual measurement varies according to the rim on which the tyre is fitted.

(2) *Nominal rim diameter*. The nominal diameter across the rim where the tyre bend seats.

(3) *Overall diameter*. The external diameter when fitted and inflated.

(4) *Section height*. The distance between the bead seat and the tread.

(5) *Rim width*. The distance between the inside faces of the rim flanges.

(6) *Load capacity*. The maximum permitted load that a tyre can carry under specified operating conditions.

What are the essential features of a cross-ply tyre?

Figure 72 clearly shows the diagonal disposition of the casing plies, placed at an angle of about 40° to the plane of rotation. This design gives a comfortable ride but due to the stiffness of the sidewalls the tread becomes distorted where it contacts the road surface, resulting in reduced traction during cornering, braking and accleration. Tyre wear is also increased.

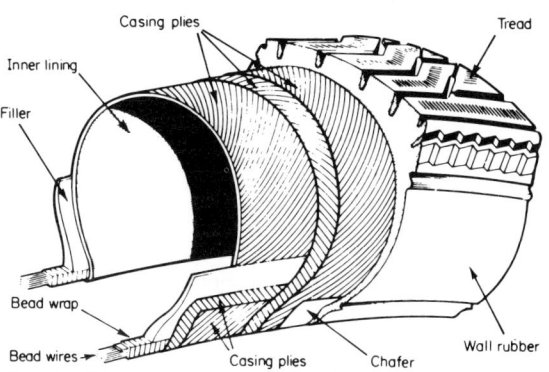

Fig. 72. *The structure of a cross-ply tyre (Pirelli)*

How does the radial-ply tyre differ from the cross-ply tyre?

The functions of cushioning and directional stability are separated in a radial tyre, by giving greater flexibility to the sidewalls whilst bracing the tread with a separate band of bracing plies. The rayon or nylon cords in each ply run from bead to bead (i.e. at 90° to the plane of rotation) which imparts great flexibility but not stability – hence the need for tread bracing layers, made from rubber-coated rayon or steel cords (Fig. 73). These bracing layers control tread movement and therefore reduce tread shuffle and wear. Radial plies produce less internal friction which in turn means cooler running and permits lower inflation pressures for increased comfort.

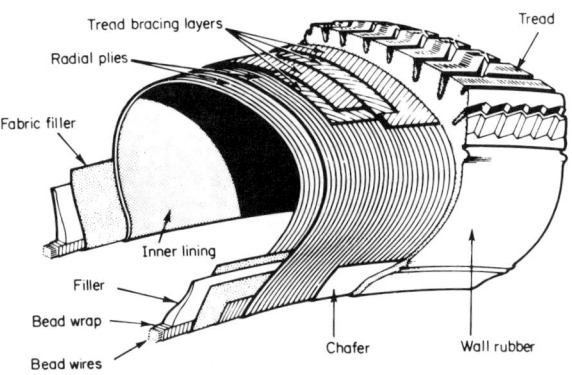

Fig. 73. *The structure of a radial-ply tyre (Pirelli)*

What are the advantages of radial-ply tyres?

(1) Improved cornering – bracing layers maintain the tread flat on road surfaces.
(2) Increased braking and accleration performance – tread bracing again.
(3) Greater comfort at higher road speeds – sidewall flexibility.
(4) Longer lasting – cooler running and braced tread.
(5) Improved fuel economy – lower rolling resistance.

114

Do radial-ply tyres have any disadvantages?

In comparison with a cross-ply tyre the radial tyre is more expensive initially, gives a harsher ride at low speeds and offers more resistance to parking manoeuvres.

Are there legal limitations on mixing cross-ply and radial-ply tyres?

Yes. These are contained in Motor Vehicles (Construction and Use) Regulations 1978, which state that it is illegal to mix radial and cross-ply tyres on the same axle, front or rear. Radials must not be fitted to the front wheels only (even if the car is front-wheel drive). Radials can be fitted to the rear wheels only, although it is recommended that the best arrangement is to have only one type all round.

Do not mix steel and textile radials on the same axle. If mixed tyres must be used, put two steel radials on the rear axle.

How is the air retained in a tubeless tyre?

The tyre casing is lined with soft rubber, which is partially self-sealing if punctured. The wheel rim must be in good condition to allow the tubeless tyre to seat efficiently to avoid loss of pressure. A tube may be fitted to overcome minor defects in the tyre or wheel rim. Tubeless tyres are usually marked as such and also are more easily balanced.

What materials are used in the construction of a tyre?

Although natural rubber has been used, most tyre manufacturers now use a variety of synthetic rubbers blended to give compounds that are both hard wearing and give good grip.

There may be as many as 20 different rubber compounds used in a tyre, but four well-known examples are: *SBR* (styrene

115

butadiene rubber), widely used in blends with natural rubber, has less bounce than natural rubber but provides good grip and softer ride; *PB* (poly-butadiene), normally blended with SBR and natural rubber to give a hard-wearing mix; *butyl*, used for tubes because it is impervious to the oxygen content of the air; and *neoprene*, a tough rubber to resist ageing, oxidation and ozonation.

In the early 1960s a new type of tread rubber called 'high mu' came into use on cars. The high mu material, in the wet, has a grip nearly as high as that of natural rubber compound in the dry.

What is the purpose of a tyre tread?

The purpose of the tread is to provide a good grip on the road surface for accelerating, braking and cornering. Although a smooth tyre would give the best grip on a dry road, owing to its greatest possible contact patch, it would be of little use in the wet because a wedge of water would build up in front of the tyre to give an effect known as aquaplaning.

How does the tread remove surface water from under the tread?

It provides drainage channels and leak paths to the rear and sides of the tyre's contact patch. At 100 km/h (60 mile/h) on a wet road, the tread pattern has to move at least 4½ litres of water out of the way every second in order to dry the surface sufficiently for good grip.

How are the sizes marked on the different types of tyre?

The size is moulded on to the sidewall of the tyre and is usually combined with other information, i.e. whether radial or crossply, together with its maximum speed rating and aspect ratio. The actual method of indicating the size depends on whether the tyre is cross-ply or radial-ply, but the information is based upon the sectional width and nominal rim diameter. Examples are:

Cross-ply   Original method: 5.20 – 13
                New method: 5.20 – S13

                5.20 = nominal section width in inches
                S   = speed symbol
                13   = nominal rim diameter in inches

Radial-ply   Original method: 165 – 13
(Michelin)   Current method: 165 SR 13 xZx

                165 = nominal section width in millimetres
                S   = speed symbol
                R   = radial construction
                13   = nominal rim diameter in inches
                xZx = tread pattern

Radial-ply   Current method: 165/70 R13 xZx 82S
(low-profile)
                165 = nominal section width in millimetres
                70   = aspect ratio
                R   = radial construction
                13   = nominal rim diameter in inches
                xZx = tread pattern
                82   = load index number
                S   = speed symbol

Why do some tyres have their rim diameter stated in millimetres?

This is a new generation of special wheel and radial-ply tyre combination, and has a slightly different manner of sidewall marking. Example:

Radial-ply   Sidewall marking: 190 65 HR 390 TRX
Michelin
TRX 70   190 = nominal section width in millimetres
series   65   = aspect ratio
                HR   = speed symbol
                390 = nominal rim diameter in millimetres
                TRX = tread pattern

These tyres differ from the other radial tyres in having ultra-low profiles or aspect ratios. Radial-ply tyres can be divided into three kinds:

(1) Standard – 82 per cent aspect ratio
(2) Low profile – 70, 60, 50 per cent aspect ratios (70, 60, 50 series)
(3) Special wheel and tyre combinations

What is meant by the term 'aspect ratio'?

Aspect ratio is simply the ratio of the inflated section height to the section width (Fig. 74) expressed as a percentage.

$$\text{Aspect ratio} = \frac{\text{section height}}{\text{section width}} \times 100 \text{ per cent}$$

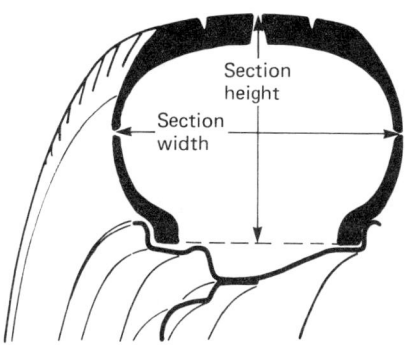

Fig. 74. *Aspect ratio*

The trend in radial-ply tyre design is to increase the sectional width while reducing the section height to improve the shape and size of the tread contact patch, thus giving a *low profile* or low aspect ratio.

How is the maximum road speed of a tyre indicated?

A system has been devised to include the maximum speed rating in the tyre size marking. A series of letters indicate particular maximum speeds; these are in the process of being updated. An extract from a table covering radial-ply tyres is given below.

Old speed marking	New speed symbol	Maximum speed mile/h	km/h
SR	P	93	150
SR	Q	99	160
SR	R	105	170
HR	T	120	190
HR	H	130	200
VR	V	over 130	over 200

How can the problem of overgearing be prevented when low profile tyres are fitted?

By fitting larger diameter wheels (Fig. 75). Larger than standard wheels are needed when fitting low profile tyres to compensate for the lower rolling diameter and to avoid overgearing.

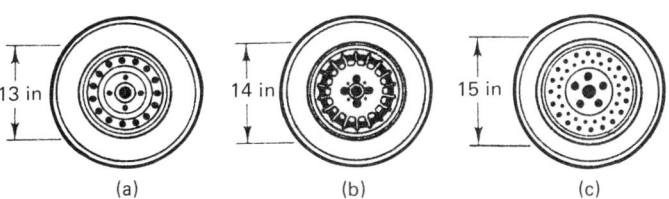

Fig. 75. Low profile tyres and larger than standard wheels: (a) 70 series, (b) 60 series, (c) 50 series

119

What are the legal requirements on tyre safety?

Tyres must be suitable for the use to which the vehicle is being put, and for use with the other tyres on it.

Tyres must be properly inflated.
Tyres must be free from breaks or cuts more than 25 mm long and deep enough to reach the body cords.
Tyres must not have any lump or bulge caused by separation or partial failure of the structure.
Tyres must not have any portion of the ply or cord structure exposed.
Tyres must have a pattern of at least 1 mm in depth over three-quarters of the tread and extending around the entire circumference.

Is wheel balancing of practical importance for driving wheels?

On FWD cars it is of prime importance. On RWD cars, static imbalance will cause 'wheel hop' or 'patter'. All wheels in balance, including the spare (for use in an emergency), is the ideal condition.

How can tyres be out-of-balance?

With the complicated structure of a modern tyre, it is impossible to obtain an even distribution of weight throughout, so that points of excess weight, under the action of centrifugal force, will cause out-of-balance effects.

These effects are caused by two forms of imbalance, static and dynamic. Static imbalance will cause 'wheel hop', 'patter' or 'tramp'. These terms explain the effect of centrifugal forces set up by the imbalance, tending to lift the wheel when the out-of-balance is at the top of the wheel, and to force it downwards when the out-of-balance is at the bottom of the wheel.

Dynamic imbalance occurs when two excess weights exist on either side of the tyre. The effect is to produce 'shimmy' in steered wheels.

How is the out-of-balance corrected?

Static and dynamic out-of-balance can be corrected by the addition of weights, at definite positions on the wheel rim. The placing of the weights and their magnitude is determined by the use of a balancing machine. Some machines balance the wheel off the car, others balance the wheel on the car.

What factors determine the life of a tyre?

Several factors are involved. Some are driver-inflicted, e.g. high-speed cornering, hard acceleration and harsh braking; others include operating temperature, road surface type and condition, wheel balance and, of course, inflation pressure.

Modern cars are capable of cornering at speeds that can reduce tyre life by several months; for example, an increase of about 10 per cent in cornering speed causes a 50 per cent increase in tread wear. Speed in itself has little effect on tyre life, provided that tyres are correctly balanced, aligned and inflated.

Why is inflation pressure so important?

Only by operating at the correct pressure can the tyres support their rated load capacity without being overloaded, and give the vehicle its correct handling performance and tyre life. Pressures should be checked at the start of a journey when the tyres are cold.

Should the tyre pressures be reduced if they are checked during a long journey?

No. Tyre pressures will increase by about $4-10\,\text{lbf/in}^2$ ($0.28-0.69$ bar) during a long journey, especially if a high speed is maintained. In fact the pressures should be increased by $2-6\,\text{lbf/in}^2$ ($0.14-0.41$ bar) for sustained high speed and when towing a caravan, etc.

Should the tyres be reduced in pressure during wet or icy weather?

No. The grip obtained will be made worse due to closing up of the tread pattern, and increased operating temperature from the excessive flexing of the widewalls will reduce tyre life.

Should wheels be changed around to equalise tread wear?

Several points should be borne in mind when deciding this issue: (1) Irregular wear on one tyre may well be due to a mechanical defect; correction of this is generally more cost-effective. (2) If all tyres are wearing evenly, nothing will be gained by interchanging. (3) If an interchange is carried out the tyre pressures may need readjustment front to rear, and wheel balance also checked.

Do new tyres require running in?

It is generally recommended to avoid high speeds for the first 100–150 miles (160–240 km).

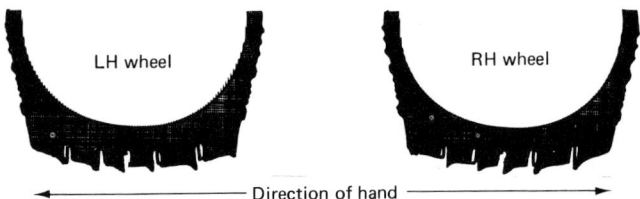

LH wheel   RH wheel

◄─────────── Direction of hand ───────────►

Fig. 76. *Wear pattern with excessive toe-in*

How can examination of the tyre tread reveal whether inflation pressure is correct or show the presence of a mechanical defect?

Excessive tread wear around the tyre circumference in particular areas is the evidence required. In the case of misaligned wheels, i.e. incorrect tracking, the tread will be feathered owing to dragging action between the tread and road surface (Fig. 76).

Index